Legal Notice

For information on bulk purchases and licensing agreements, please email

support@SATPrepGet800.com

ISBN-13: 978-1-951619-05-3

This is the Solution Guide to the book
"Set Theory for Pre-Beginners."

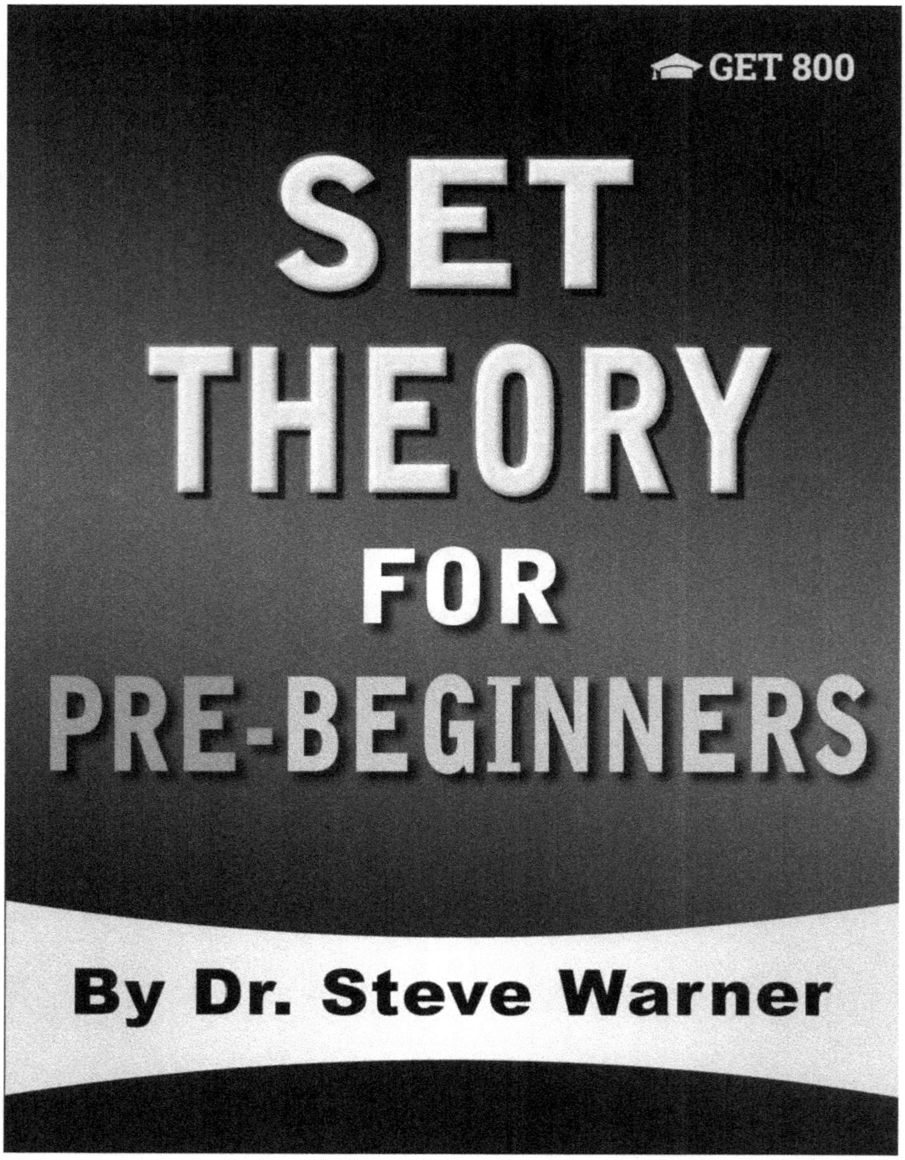

Also Available from Dr. Steve Warner

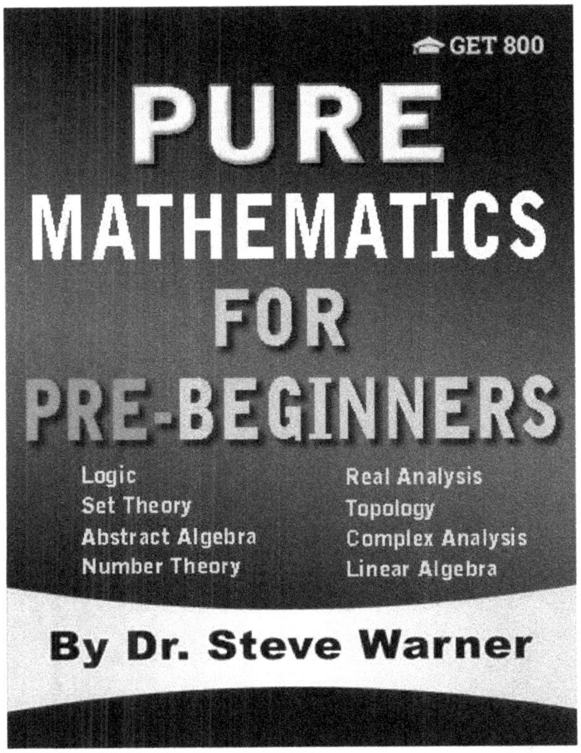

CONNECT WITH DR. STEVE WARNER

www.facebook.com/SATPrepGet800

www.youtube.com/TheSATMathPrep

www.twitter.com/SATPrepGet800

www.linkedin.com/in/DrSteveWarner

www.pinterest.com/SATPrepGet800

Also Available from Dr. Steve Warner

CONNECT WITH DR. STEVE WARNER

www.facebook.com/SATPrepGet800

www.youtube.com/TheSATMathPrep

www.twitter.com/SATPrepGet800

www.linkedin.com/in/DrSteveWarner

www.pinterest.com/SATPrepGet800

Set Theory for Pre-Beginners

Soluton Guide

Dr. Steve Warner

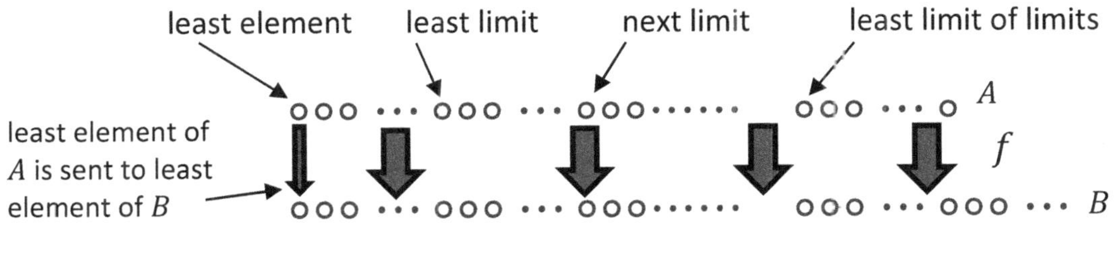

Table of Contents

Problem Set 1

LEVEL 1

Determine whether each of the following statements is true or false:

1. $t \in \{t\}$

$\{t\}$ has exactly 1 element, namely t. So, $t \in \{t\}$ is **true**.

2. $4 \in \{0, 2, 4, 6\}$

$\{0, 2, 4, 6\}$ has exactly 4 elements, namely 0, 2, 4, and 6. In particular, $4 \in \{0, 2, 4, 6\}$ is **true**.

3. $-7 \in \{7\}$

$\{7\}$ has exactly 1 element, namely 7. So, $-7 \notin \{7\}$. Therefore, $-7 \in \{7\}$ is **false**.

4. $0 \in \mathbb{Z}$

$\mathbb{Z} = \{\dots, -4, -3, -2, -1, 0, 1, 2, 3, 4, \dots\}$. In particular, $0 \in \mathbb{Z}$ is **true**.

5. $-26 \in \mathbb{N}$

$\mathbb{N} = \{0, 1, 2, 3, \dots\}$. Therefore, $-26 \in \mathbb{N}$ is **false**.

6. $\frac{7}{2} \in \mathbb{Q}$

Since $7, 2 \in \mathbb{Z}$ and $2 \neq 0$, $\frac{7}{2} \in \mathbb{Q}$ is **true**.

7. $\emptyset \subseteq \{1, 2, 3\}$

The empty set is a subset of every set. So, $\emptyset \subseteq \{1, 2, 3\}$ is **true**.

8. $\{\square\} \subseteq \{\square, \Delta\}$

The only element of $\{\square\}$ is $\square$. Since $\square$ is also an element of $\{\square, \Delta\}$, $\{\square\} \subseteq \{\square, \Delta\}$ is **true**.

9. $\{x, y, z, w\} \subset \{x, y, z, w\}$

No set is a **proper** subset of itself. So, $\{x, y, z, w\} \subset \{x, y, z, w\}$ is **false**.

10. $\{0, 1, \{2, 3\}\} \subseteq \{0, 1, 2, 3\}$

$\{2, 3\} \in \{0, 1, \{2,3\}\}$, but $\{2, 3\} \notin \{0, 1, 2, 3\}$. So, $\{0, 1, \{2,3\}\} \subseteq \{0, 1, 2, 3\}$ is **false**.

Determine the cardinality of each of the following sets:

11. {hammer, screwdriver, saw}

$|\{\text{hammer}, \text{screwdriver}, \text{saw}\}| = \mathbf{3}$.

12. $\{4, 16, 27, 46, 59, 201\}$

$|\{4, 16, 27, 46, 59, 201\}| = \mathbf{6}$.

13. $\{1, 2, \dots, 236\}$

$|\{1, 2, \dots, 236\}| = \mathbf{236}$.

14. $\left\{\frac{1}{2}, \frac{1}{3}, \dots, \frac{1}{15}\right\}$

$\left|\left\{\frac{1}{2}, \frac{1}{3}, \dots, \frac{1}{15}\right\}\right| = \mathbf{14}$.

15. $\emptyset$

$|\emptyset| = \mathbf{0}$.

Provide an example of a set X with the given properties:

16. (i) $X \subset \mathbb{Z}$ (X is a *proper* subset of $\mathbb{Z}$); (ii) X is infinite; (iii) X contains both positive and negative integers; (iv) X contains both even and odd integers.

One example of such a set X is $X = \{\dots, -\mathbf{6}, -\mathbf{4}, -\mathbf{2}, \mathbf{1}, \mathbf{3}, \mathbf{5}, \dots\}$.

17. (i) $X \subset \mathbb{R}$ (X is a *proper* subset of $\mathbb{R}$); (ii) X contains both rational and irrational numbers.

One example of such a set X is $X = \{\mathbf{0.1}, \mathbf{0.101001000100001}\dots\}$.

18. (i) $X \subset \mathbb{C}$ (X is a *proper* subset of $\mathbb{C}$); (ii) X is infinite; (iii) X contains real numbers; and (iv) X contains complex numbers that are not real.

One example of such a set X is $X = \{\boldsymbol{a} + \boldsymbol{bi} \mid \boldsymbol{a} \in \mathbb{R} \text{ and } (\boldsymbol{b} = \mathbf{0} \text{ or } \boldsymbol{b} = \mathbf{1})\}$.

For each pair of sets A and B below, determine if $A \subseteq B$, $B \subseteq A$, both, or neither.

19. $A = 2\mathbb{N}$, $B = \mathbb{Z}$

$\boldsymbol{A} \subseteq \boldsymbol{B}$

20. $A = \{0, 1\}$, $B = \{1, 0\}$

$\boldsymbol{A} = \boldsymbol{B}$ **(Both)**

21. $A = \emptyset$, $B = \{\}$

$A = B$ **(Both)**

22. $A = \{0, 1, 2, 3, 4\}$, $B = \{1, 3\}$

$B \subseteq A$

23. $A = \{a\}$, $B = \{b\}$

Neither (unless $a = b$, in which case $A = B$ and the answer is **Both**)

LEVEL 2

Determine whether each of the following statements is true or false:

24. $a \in \emptyset$

The empty set has no elements. So, $x \in \emptyset$ is false for any x. In particular, $a \in \emptyset$ is **false**.

25. $\emptyset \in \emptyset$

The empty set has no elements. So, $x \in \emptyset$ is false for any x. In particular, $\emptyset \in \emptyset$ is **false**.

26. $\emptyset \in \{\emptyset, \{\emptyset\}\}$

The set $\{\emptyset, \{\emptyset\}\}$ has exactly 2 elements, namely $\emptyset$ and $\{\emptyset\}$. In particular, $\emptyset \in \{\emptyset\}$ is **true**.

27. $\{\emptyset\} \in \emptyset$

The empty set has no elements. So, $x \in \emptyset$ is false for any x. In particular, $\{\emptyset\} \in \emptyset$ is **false**.

28. $\{\emptyset\} \in \{\emptyset\}$

The set $\{\emptyset\}$ has 1 element, namely $\emptyset$. Since $\{\emptyset\} \neq \emptyset$, $\{\emptyset\} \in \{\emptyset\}$ is **false**.

29. $7 \in 2\mathbb{N}$

$2\mathbb{N} = \{0, 2, 4, 6, 8, 10, 12, 14, \ldots\}$. In particular, $7 \in 2\mathbb{N}$ is **false**.

30. $\emptyset \subseteq \emptyset$

The empty set is a subset of every set. So, $\emptyset \subseteq X$ is true for any X. In particular, $\emptyset \subseteq \emptyset$ is **true**. (This can also be done by using the fact that every set is a subset of itself.)

31. $\emptyset \subseteq \{\emptyset\}$

Again, (as in Problem 30), $\emptyset \subseteq X$ is true for any X. In particular, $\emptyset \subseteq \{\emptyset\}$ is **true**.

32. $\{\emptyset\} \subseteq \emptyset$

The only subset of $\emptyset$ is $\emptyset$. So, $\{\emptyset\} \subseteq \emptyset$ is **false**.

33. $\{\emptyset\} \subseteq \{\emptyset\}$

Every set is a subset of itself. So, $\{\emptyset\} \subseteq \{\emptyset\}$ is **true**.

Determine the cardinality of each of the following sets:

34. $\{a, a, b, c, c, c, d, d\}$

$\{a, a, b, c, c, c, d, d\} = \{a, b, c, d\}$. Therefore, $|\{a, a, b, c, c, c, d, d\}| = |\{a, b, c, d\}| = \mathbf{4}$.

35. $\{\{x, y\}, \{z, u, v\}, \{w\}\}$

$\{\{x, y\}, \{z, u, v\}, \{w\}\}$ consists of the 3 elements $\{x, y\}$, $\{z, u, v\}$, and $\{w\}$. Therefore, we have $|\{\{x, y\}, \{z, u, v\}, \{w\}\}| = \mathbf{3}$.

36. $\{5, 6, 7, \dots, 4322, 4323\}$

$|\{5, 6, 7, \dots, 4322, 4323\}| = 4323 - 5 + 1 = \mathbf{4319}$ (notice that we used the fence-post formula here).

Determine if each of the following real numbers is rational or irrational:

37. $5.\overline{7}$

rational (repeating decimal)

38. $-72.8101214161820222242628\dots$

irrational

39. $463.654321543215432154321\dots$

rational ($463.654321543215432154321\dots = 463.6\overline{54321}$, repeating decimal)

40. 0

rational (terminating decimal)

LEVEL 3

Use set-builder notation to describe each of the following sets.

41. $\{2, 4, 6, 8, 10, 12, 14, 16, 18, 20, 22\}$

$\{2, 4, 6, 8, 10, 12, 14, 16, 18, 20, 22\} = \{\boldsymbol{n} \in \mathbb{N} \mid \boldsymbol{n} \textbf{ is even and } \mathbf{2} \le \boldsymbol{n} \le \mathbf{22}\}$

42. $2\mathbb{N}$

$2\mathbb{N} = \{\boldsymbol{n} \in \mathbb{N} \mid \boldsymbol{n} \textbf{ is even}\}$

43. $\mathbb{Z}^-$

$\mathbb{Z}^- = \{\boldsymbol{n} \in \mathbb{Z} \mid \boldsymbol{n} \textbf{ is negative}\}$

Determine the cardinality of each of the following sets:

44. $\{\{\{x, y, z\}\}\}$

The only element of $\{\{\{x, y, z\}\}\}$ is $\{\{x, y, z\}\}$. So, $|\{\{\{x, y, z\}\}\}| = \mathbf{1}$.

45. $\{\{0, 1\}, 0, \{0\}, \{0, \{0, 1, 2\}\}\}$

The elements of $\{\{0, 1\}, 0, \{0\}, \{0, \{0, 1, 2\}\}\}$ are $\{0, 1\}$, 0, $\{0\}$, and $\{0, \{0, 1, 2\}\}$. So, we see that $|\{\{0, 1\}, 0, \{0\}, \{0, \{0, 1, 2\}\}\}| = \mathbf{4}$.

46. $\{a, \{a\}, \{a, a\}, \{a, a, a, a\}, \{a, a, \{a\}\}, \{a, \{a\}, \{a\}\}\}$

$$\begin{aligned}
&\{a, \{a\}, \{a, a\}, \{a, a, a, a\}, \{a, a, \{a\}\}, \{a, \{a\}, \{a\}\}\} \\
&= \{a, \{a\}, \{a\}, \{a\}, \{a, \{a\}\}, \{a, \{a\}\}\} \\
&= \{a, \{a\}, \{a, \{a\}\}\}.
\end{aligned}$$

So, $|\{a, \{a\}, \{a, a\}, \{a, a, a, a\}, \{a, a, \{a\}\}, \{a, \{a\}, \{a\}\}\}| = |\{a, \{a\}, \{a, \{a\}\}\}| = \mathbf{3}$.

For each set X, determine $|\mathcal{P}(X)|$

47. $X = \{0, 1, 2, 3, 4, 5, 6\}$

Since $|X| = 7$, $|\mathcal{P}(X)| = 2^7 = \mathbf{128}$.

48. $X = \{\emptyset, \{\emptyset\}, \{\emptyset, \{\emptyset\}\}\}$

Since $|X| = 3$, $|\mathcal{P}(X)| = 2^3 = \mathbf{8}$.

49. $X = \{26, 27, 28, \dots, 203, 204\}$

Since $|X| = 204 - 26 + 1 = 179$, $|\mathcal{P}(X)| = \mathbf{2^{179}}$.

Level 4

Determine whether each of the following statements is true or false:

50. $x \in \{x, \{y\}\}$

The set $\{x, \{y\}\}$ has exactly 2 elements, namely x and $\{y\}$. So, $x \in \{x, \{y\}\}$ is **true**.

51. $3 \in \{2k \mid k = 1, 2, 3, 4\}$

$\{2k \mid k = 1, 2, 3, 4\} = \{2, 4, 6, 8\}$. So, $3 \notin \{2k \mid k = 1, 2, 3, 4\}$. Therefore, $3 \in \{2k \mid k = 1, 2, 3, 4\}$ is **false**.

52. $\{0\} \in \{0, 1\}$

The set $\{0, 1\}$ has exactly 2 elements, namely 0 and 1. So, $\{0\} \in \{0, 1\}$ is **false**.

53. $\{1\} \in \{\{1\}, x, 2, y\}$

The set $\{\{1\}, x, 2, y\}$ has exactly 4 elements, namely $\{1\}$, x, 2 , and y. So, $\{1\} \in \{\{1\}, x, 2, y\}$ is **true**.

54. $\emptyset \in \{\{\emptyset\}\}$

The set $\{\{\emptyset\}\}$ has exactly 1 element, namely $\{\emptyset\}$. Since $\emptyset$ is not equal to $\{\emptyset\}$, $\emptyset \in \{\{\emptyset\}\}$ is **false**.

55. $\{\{\emptyset\}\} \in \emptyset$

The empty set has no elements. So, $x \in \emptyset$ is false for any x. In particular, $\{\{\emptyset\}\} \in \emptyset$ is **false**.

Compute the power set of each of the following sets:

56. $\emptyset$

$\mathcal{P}(\emptyset) = \{\emptyset\}$

57. $\{\emptyset\}$

$\mathcal{P}(\{0\}) = \{\emptyset, \{\emptyset\}\}$

58. {lion, tiger, jaguar}

$\mathcal{P}(\{\text{lion}, \text{tiger}, \text{jaguar}\}) =$

$$\{\emptyset, \{\text{lion}\}, \{\text{tiger}\}, \{\text{jaguar}\}, \{\text{lion}, \text{tiger}\}, \{\text{lion}, \text{jaguar}\}, \{\text{tiger}, \text{jaguar}\}, \{\text{lion}, \text{tiger}, \text{jaguar}\}\}$$

59. $\{\emptyset, \{\emptyset\}\}$

$\mathcal{P}(\{\emptyset, \{\emptyset\}\}) = \{\emptyset, \{\emptyset\}, \{\{\emptyset\}\}, \{\emptyset, \{\emptyset\}\}\}$

60. $\{\{\emptyset\}\}$

$$\mathcal{P}(\{\{\emptyset\}\}) = \{\emptyset, \{\{\emptyset\}\}\}$$

61. $\{\emptyset, \{\emptyset\}, \{\emptyset, \{\emptyset\}\}\}$

$$\mathcal{P}\left(\{\emptyset, \{\emptyset\}, \{\emptyset, \{\emptyset\}\}\}\right) =$$

$$\{\emptyset, \{\emptyset\}, \{\{\emptyset\}\}, \{\{\emptyset, \{\emptyset\}\}\}, \{\emptyset, \{\emptyset\}\}, \{\emptyset, \{\emptyset, \{\emptyset\}\}\}, \{\{\emptyset\}, \{\emptyset, \{\emptyset\}\}\}, \{\emptyset, \{\emptyset\}, \{\emptyset, \{\emptyset\}\}\}\}$$

A **relation** describes a relationship between objects. For example, the relation = describes the relationship "is equal to." Two other relations we have seen are $\in$ (the membership relation) and $\subseteq$ (the subset relation). A relation R is **reflexive** if for all x, we have xRx. A relation R is **symmetric** if for all x,y, we have $xRy \to yRx$. A relation R is **transitive** if for all x,y, z, we have $(xRy \wedge yRz) \to xRz$. For example, the relation "=" is reflexive, symmetric, and transitive because for all x, we have $x = x$, for all x,y, we have $x = y \to y = x$, and for all x,y, z, we have $(x = y \wedge y = z) \to x = z$.

62. Is $\subseteq$ reflexive?

Yes. Every set is a subset of itself ($A \subseteq A$).

63. Is $\in$ reflexive?

No. Since the empty set has no elements, $\emptyset \notin \emptyset$. This **counterexample** shows that $\in$ is not reflexive.

64. Is $\subseteq$ symmetric?

No. $\{1\} \subseteq \{1, 2\}$, but $\{1, 2\} \nsubseteq \{1\}$. This **counterexample** shows that $\subseteq$ is not symmetric.

65. Is $\in$ symmetric?

No. $\emptyset \in \{\emptyset\}$, but $\{\emptyset\} \notin \emptyset$. This **counterexample** shows that $\in$ is not symmetric.

66. Is $\subseteq$ transitive?

Yes. If $A \subseteq B$ and $B \subseteq C$, then $A \subseteq C$.

67. Is $\in$ transitive?

No. $\emptyset \in \{\emptyset\} \in \{\{\emptyset\}\}$, but $\emptyset \notin \{\{\emptyset\}\}$. This **counterexample** shows that $\in$ is not transitive.

Explicitly write down each of the following natural numbers using only set brackets and the empty set.

68. 5

$$5 = \{0, 1, 2, 3, 4\} = \{\emptyset, \{\emptyset\}, \{\emptyset, \{\emptyset\}\}, \{\emptyset, \{\emptyset\}, \{\emptyset, \{\emptyset\}\}\}, \{\emptyset, \{\emptyset\}, \{\emptyset, \{\emptyset\}\}, \{\emptyset, \{\emptyset\}, \{\emptyset, \{\emptyset\}\}\}\}\}$$

69. 6

$$6 = \{0, 1, 2, 3, 4, 5\} =$$

$$\Big\{\emptyset, \{\emptyset\}, \{\emptyset, \{\emptyset\}\}, \{\emptyset, \{\emptyset\}, \{\emptyset, \{\emptyset\}\}\}, \{\emptyset, \{\emptyset\}, \{\emptyset, \{\emptyset\}\}, \{\emptyset, \{\emptyset\}, \{\emptyset, \{\emptyset\}\}\}\}, \{\emptyset, \{\emptyset\}, \{\emptyset, \{\emptyset\}\}, \{\emptyset, \{\emptyset\}, \{\emptyset, \{\emptyset\}\}\}, \{\emptyset, \{\emptyset\}, \{\emptyset, \{\emptyset\}\}, \{\emptyset, \{\emptyset\}, \{\emptyset, \{\emptyset\}\}\}\}\}\Big\}$$

LEVEL 5

We say that a set A is **transitive** if every element of A is a subset of A. Determine if each of the following sets is transitive:

70. $\emptyset$

Since $\emptyset$ has no elements, **$\emptyset$ is transitive**. (The statement "$x \in \emptyset \to x \subseteq \emptyset$" is true simply because "$x \in \emptyset$" is always false. In this case, we say that the statement is **vacuously true**.)

71. $\{\emptyset\}$

The only element of $\{\emptyset\}$ is $\emptyset$, and $\emptyset \subseteq \{\emptyset\}$ is true. So, **$\{\emptyset\}$ is transitive**.

72. $\{\{\emptyset\}\}$

$\{\emptyset\} \in \{\{\emptyset\}\}$ and $\emptyset \in \{\emptyset\}$, but $\emptyset \notin \{\{\emptyset\}\}$. So, **$\{\{\emptyset\}\}$ is not transitive**.

73. $\{\emptyset, \{\emptyset\}\}$

$\{\emptyset, \{\emptyset\}\}$ has 2 elements, namely $\emptyset$ and $\{\emptyset\}$. Both sets are subsets of $\{\emptyset, \{\emptyset\}\}$. It follows that **$\{\emptyset, \{\emptyset\}\}$ is transitive**.

74. $\{\emptyset, \{\emptyset\}, \{\{\emptyset\}\}\}$

$\{\emptyset, \{\emptyset\}, \{\{\emptyset\}\}\}$ has 3 elements, namely $\emptyset$, $\{\emptyset\}$, and $\{\{\emptyset\}\}$. All three of these sets are subsets of $\{\emptyset, \{\emptyset\}, \{\{\emptyset\}\}\}$. It follows that **$\{\emptyset, \{\emptyset\}, \{\{\emptyset\}\}\}$ is transitive**.

75. $\{\{\emptyset\}, \{\emptyset, \{\emptyset\}\}\}$

$\{\emptyset\} \in \{\{\emptyset\}, \{\emptyset, \{\emptyset\}\}\}$ and $\emptyset \in \{\emptyset\}$, but $\emptyset \notin \{\{\emptyset\}, \{\emptyset, \{\emptyset\}\}\}$. So, **$\{\{\emptyset\}, \{\emptyset, \{\emptyset\}\}\}$ is not transitive**.

76. Assuming that A is transitive, is $\mathcal{P}(A)$ transitive?

Yes. To see this, let $x \in \mathcal{P}(A)$, and let $y \in x$. Since $x \in \mathcal{P}(A)$, $x \subseteq A$. Since $y \in x$ and $x \subseteq A$, $y \in A$. Since A is transitive and $y \in A$, $y \subseteq A$. So, $y \in \mathcal{P}(A)$. Since $y \in x$ was arbitrary, $x \subseteq \mathcal{P}(A)$. Therefore, every element of $\mathcal{P}(A)$ is a subset of $\mathcal{P}(A)$. So, $\mathcal{P}(A)$ is transitive.

Let A and B be sets with $B \subseteq A$. Determine if the following are true or false.

77. $B \in \mathcal{P}(A)$

True. This follows from the definition of $\mathcal{P}(A)$.

78. $B \subseteq \mathcal{P}(A)$

False. Let $A = \{a, b\}$ and $B = \{a\}$. Then $\mathcal{P}(A) = \{\emptyset, \{a\}, \{b\}, \{a, b\}\}$. Now, $a \in B$, but $a \notin \mathcal{P}(A)$. Therefore, $B \nsubseteq \mathcal{P}(A)$.

79. $\mathcal{P}(B) \in \mathcal{P}(A)$

False. Let $A = \{0, 1\}$ and $B = \{1\}$. Then $\mathcal{P}(A) = \{\emptyset, \{0\}, \{1\}, \{0, 1\}\}$ and $\mathcal{P}(B) = \{\emptyset, \{1\}\} \notin \mathcal{P}(A)$.

80. $\mathcal{P}(B) \subseteq \mathcal{P}(A)$

True. Let A and B be sets with $B \subseteq A$ and let $X \in \mathcal{P}(B)$. Then $X \subseteq B$ Since $X \subseteq B$ and $B \subseteq A$, by the transitivity of $\subseteq$, $X \subseteq A$. So, $X \in \mathcal{P}(A)$. Since $X \in \mathcal{P}(B)$ was arbitrary, $\mathcal{P}(B) \subseteq \mathcal{P}(A)$.

Problem Set 2

LEVEL 1

Let $A = \{x, y, z, w\}$ and $B = \{s, t, y, w\}$. Determine each of the following:

1. $A \cup B$

$A \cup B = \{\boldsymbol{x}, \boldsymbol{y}, \boldsymbol{z}, \boldsymbol{w}, \boldsymbol{s}, \boldsymbol{t}\}$

2. $A \cap B$

$A \cap B = \{\boldsymbol{y}, \boldsymbol{w}\}$

3. $A \setminus B$

$A \setminus B = \{\boldsymbol{x}, \boldsymbol{z}\}$

4. $B \setminus A$

$B \setminus A = \{\boldsymbol{s}, \boldsymbol{t}\}$

5. $A \Delta B$

$A \Delta B = (A \setminus B) \cup (B \setminus A) = \{x, z\} \cup \{s, t\} = \{\boldsymbol{x}, \boldsymbol{z}, \boldsymbol{s}, \boldsymbol{t}\}$

Determine if each of the following sets is an interval.

6. $A = \{x \in \mathbb{R} \mid 12 \leq x \leq 15\}$

Yes

7. $B = \{x \in \mathbb{R} \mid x < -103\}$

Yes

8. $C = \{x \in \mathbb{Q} \mid x < -103\}$

No because there are irrational numbers between any two rational numbers.

9. $D = \mathbb{Q}^-$

No because there are irrational numbers between any two rational numbers.

10. $E = \mathbb{R}^+$

Yes

11. $F = \{x \in \mathbb{R} \mid x \geq -16\}$

Yes

12. $G = \{x \in \mathbb{R} \mid 0 \leq x < 999\}$

Yes

13. $H = \mathbb{R} \setminus \{0\}$

No: For example, $-1 < 0 < 1$, $-1, 1 \in \mathbb{R} \setminus \{0\}$, but $0 \notin \mathbb{R} \setminus \{0\}$.

Sketch the graph of each of the following:

14. $\mathbb{R}$

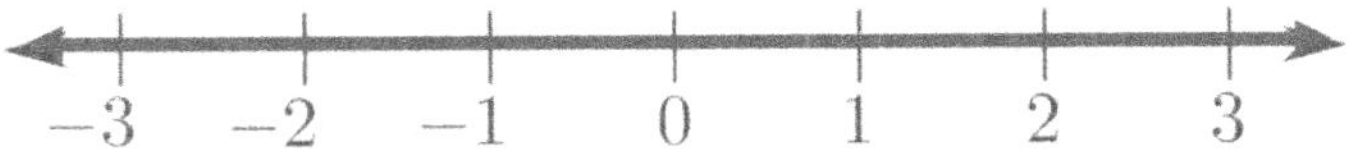

15. $\mathbb{R}^+$

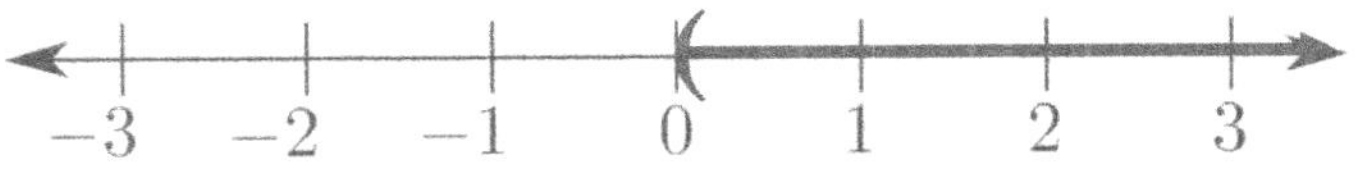

16. $\{-1, 1\}$

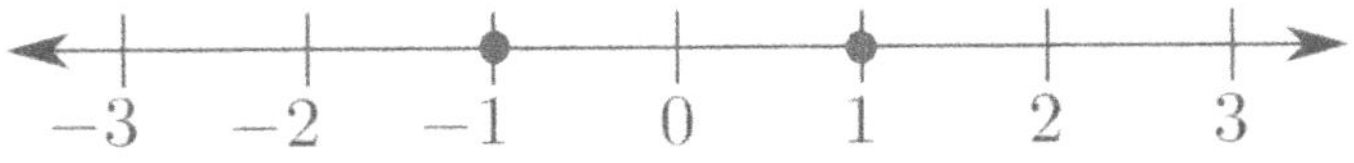

17. $(-1, 1)$

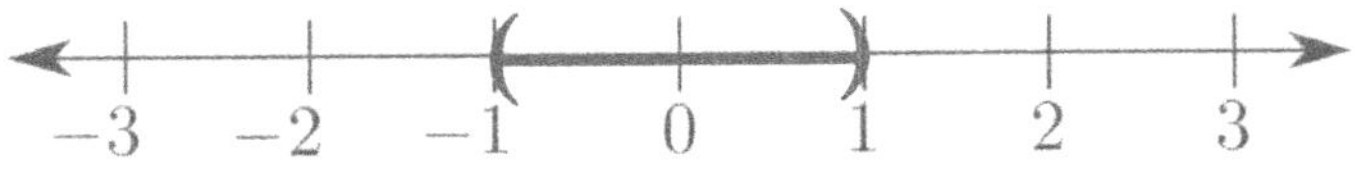

18. $[-1, \infty)$

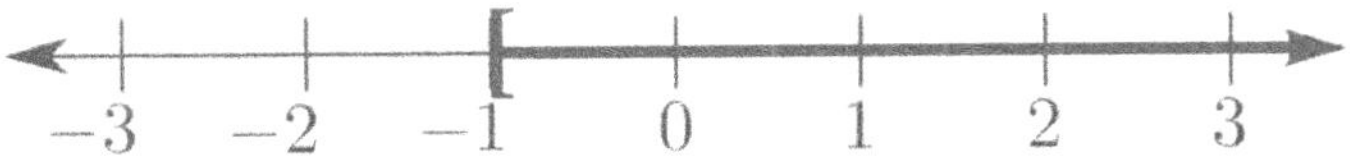

19. $\mathbb{N}$

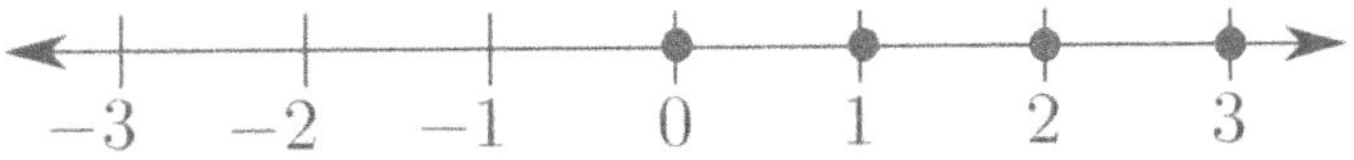

20. $\mathbb{Z}$

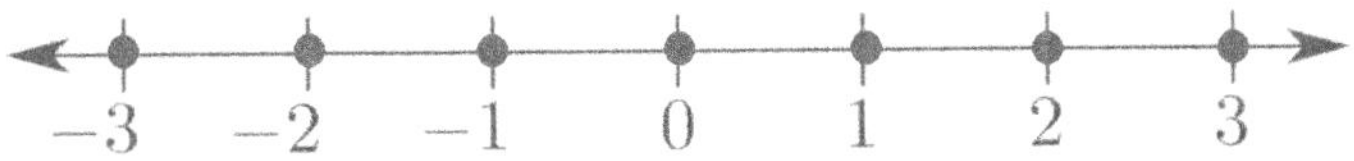

21. $(-\infty,-1)$

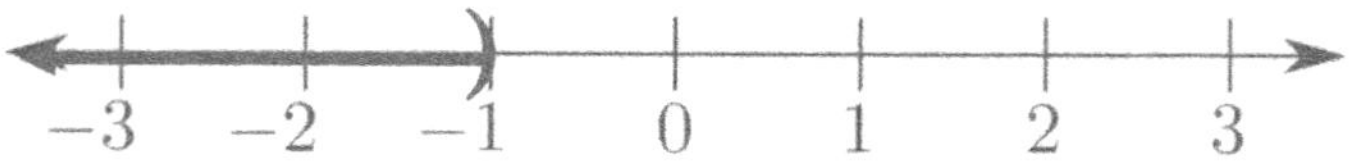

22. $(1,2]$

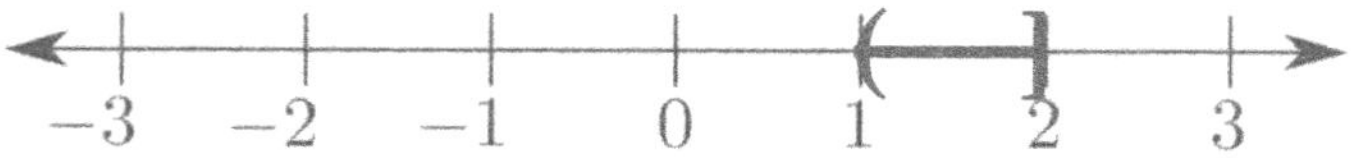

LEVEL 2

Let $A = \{\emptyset,\{\emptyset,\{\emptyset\}\}\}$ and $B = \{\emptyset,\{\emptyset\}\}$. Compute each of the following:

23. $A \cup B$

$A \cup B = \{\emptyset,\{\emptyset\},\{\emptyset,\{\emptyset\}\}\}$

24. $A \cap B$

$A \cap B = \{\emptyset\}$

25. $A \setminus B$

$A \setminus B = \{\{\emptyset,\{\emptyset\}\}\}$

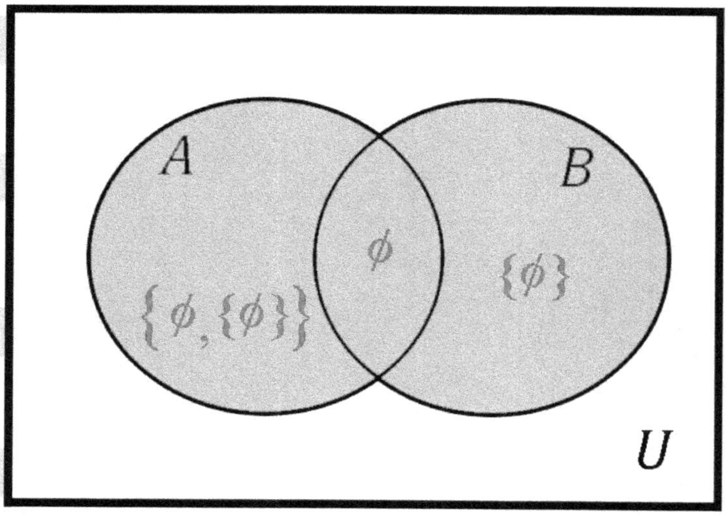

26. $B \setminus A$

$B \setminus A = \{\{\emptyset\}\}$

27. $A \,\Delta\, B$

$A \,\Delta\, B = \{\{\emptyset,\{\emptyset\}\}\} \cup \{\{\emptyset\}\} = \{\{\emptyset\},\{\emptyset,\{\emptyset\}\}\}$

Let $A = (-17,6)$ and $B = (-1,17)$. Compute each of the following:

28. $A \cup B$

$(-\mathbf{17},\mathbf{17})$

29. $A \cap B$

$(-\mathbf{1},\mathbf{6})$

30. $A \setminus B$

$(-\mathbf{17}, -\mathbf{1}]$

31. $B \setminus A$

$[\mathbf{6}, \mathbf{17})$

32. $A \,\Delta\, B$

$(-\mathbf{17}, -\mathbf{1}] \cup [\mathbf{6}, \mathbf{17})$

Let $A = [14, \infty)$ and $B = (-\infty, 15)$. Compute each of the following:

33. $A \cup B$

$(-\infty, \infty)$

34. $A \cap B$

$[\mathbf{14}, \mathbf{15})$

35. $A \setminus B$

$[\mathbf{15}, \infty)$

36. $B \setminus A$

$(-\infty, \mathbf{14})$

37. $A \,\Delta\, B$

$(-\infty, \mathbf{14}) \cup [\mathbf{15}, \infty)$

LEVEL 3

Let A, B, and C be sets, let $X = (A \setminus B) \setminus C$, and let $Y = A \setminus (B \setminus C)$.

38. Draw Venn Diagrams for X and Y.

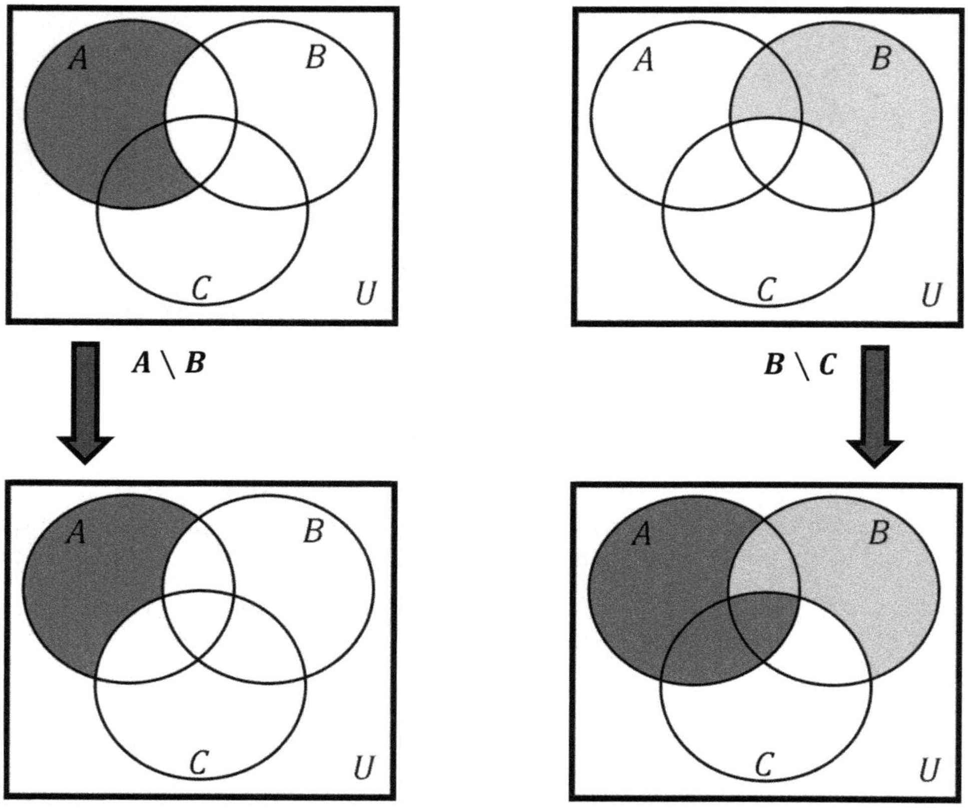

$X = (A \setminus B) \setminus C$ $\qquad$ $Y = A \setminus (B \setminus C)$

39. Is $X \subseteq Y$?

Yes

40. Is $Y \subseteq X$?

No

41. Is $X = Y$?

No

For each of the following, compute $\bigcup \boldsymbol{X}$ and $\bigcap \boldsymbol{X}$.

42. $\boldsymbol{X} = \{\{0,1,2\},\{1,2,3\},\{2,3,4\}\}$

$\bigcup \boldsymbol{X} = \{\mathbf{0},\mathbf{1},\mathbf{2},\mathbf{3},\mathbf{4}\}$ $\quad \bigcap \boldsymbol{X} = \{\mathbf{2}\}$

43. $\boldsymbol{X} = \{\mathbb{N}, \mathbb{Z}, \mathbb{Q}, \mathbb{R}\}$

$\bigcup \boldsymbol{X} = \mathbb{R}$ $\qquad \bigcap \boldsymbol{X} = \mathbb{N}$

44. $\boldsymbol{X} = \{(0,10),(1,11),(2,12),(3,13),(4,14)\}$

$\bigcup \boldsymbol{X} = (\mathbf{0},\mathbf{14})$ $\bigcap \boldsymbol{X} = (\mathbf{4},\mathbf{10})$

45. $\boldsymbol{X} = \{(-\infty,20),[-5,17),(4,100]\}$

$\bigcup \boldsymbol{X} = (-\infty,\mathbf{100}]$ $\bigcap \boldsymbol{X} = (\mathbf{4},\mathbf{17})$

46. $\boldsymbol{X} = \{(0,1],(1,2],(2,3],(3,4]\}$

$\bigcup \boldsymbol{X} = (\mathbf{0},\mathbf{4}]$ $\bigcap \boldsymbol{X} = \emptyset$

Find the natural number that each of the following expressions is equal to.

47. 7^+

$7^+ = 7 \cup \{7\} = \{0,1,2,3,4,5,6\} \cup \{7\} = \{0,1,2,3,4,5,6,7\} = \mathbf{8}$

48. $12 \cup \{12\}$

$12 \cup \{12\} = \{0,1,2,3,4,5,6,7,8,9,10,11\} \cup \{12\} = \{0,1,2,3,4,5,6,7,8,9,10,11,12\} = \mathbf{13}$

49. $8 \cap 3$

$8 \cap 3 = \{0,1,2,3,4,5,6,7\} \cap \{0,1,2\} = \{0,1,2\} = \mathbf{3}$

LEVEL 4

For each of the following, compute $\bigcup \boldsymbol{X}$ and $\bigcap \boldsymbol{X}$.

50. $\boldsymbol{X} = \{(0,q] \mid q \in \mathbb{Q}^+\}$

$\bigcup \boldsymbol{X} = (\mathbf{0},\infty)$ $\bigcap \boldsymbol{X} = \emptyset$

51. $\boldsymbol{X} = \left\{\left(\frac{1}{n},1\right) \middle| n \in \mathbb{Z}^+\right\}$

$\bigcup \boldsymbol{X} = (\mathbf{0},\mathbf{1})$ $\bigcap \boldsymbol{X} = \emptyset$

52. $\boldsymbol{X} = \{(n,n+3) \mid n \in \mathbb{Z}\}$

$\bigcup \boldsymbol{X} = \mathbb{R}$ $\bigcap \boldsymbol{X} = \emptyset$

53. $\boldsymbol{X} = \left\{\left(0,1+\frac{1}{n}\right) \mid n \in \mathbb{Z}^+\right\}$

$\bigcup \boldsymbol{X} = (\mathbf{0},\mathbf{2})$ $\bigcap \boldsymbol{X} = (\mathbf{0},\mathbf{1}]$

54. $\boldsymbol{X} = \left\{\left(-\infty, \frac{1}{n}\right] \middle| n \in \mathbb{Z}^+\right\}$

$\bigcup \boldsymbol{X} = (-\infty, \mathbf{1}]$ $\bigcap \boldsymbol{X} = (-\infty, \mathbf{0}]$

If $\boldsymbol{X}$ is a nonempty set of sets, we say that $\boldsymbol{X}$ is **disjoint** if $\bigcap \boldsymbol{X} = \emptyset$. We say that $\boldsymbol{X}$ is **pairwise disjoint** if for all $A, B \in \boldsymbol{X}$ with $A \neq B$, A and B are disjoint. For each of the following, determine if $\boldsymbol{X}$ is disjoint, pairwise disjoint, both, or neither.

55. $\{(n, n+1) \mid n \in \mathbb{Z}\}$

Both

56. $\{(n, n+1] \mid n \in \mathbb{Z}\}$

Both

57. $\left\{\left(\frac{1}{n+1}, \frac{1}{n}\right) \mid n \in \mathbb{Z}^+\right\}$

Both

58. $\{\mathbb{Q}\}$

Neither

59. $\{\mathbb{Q}^-, 2\mathbb{N}, \{n \in \mathbb{N} \mid n \text{ is a prime number greater than } 2\}\}$

Both

LEVEL 5

Let A and B be sets with $B \subseteq A$. Determine if the following are true or false.

60. $A \cap B = A$

False. Let $A = \{0, 1\}$ and $B = \{1\}$. Then $A \cap B = \{1\} \neq A$.

61. $A \setminus B \subseteq A$

True. Let $x \in A \setminus B$. Then $x \in A$ and $x \notin B$. In particular, $x \in A$. Since $x \in A \setminus B$ was arbitrary, $A \setminus B \subseteq A$.

Let $\boldsymbol{X}$ be a nonempty set of sets. Verify each of the following:

62. For all $A \in \boldsymbol{X}$, $A \subseteq \bigcup \boldsymbol{X}$.

Let $A \in \boldsymbol{X}$ and let $x \in A$. Then there is $B \in \boldsymbol{X}$ such that $x \in B$ (namely A). So, $x \in \bigcup \boldsymbol{X}$. Since x was an arbitrary element of A, we have shown that $A \subseteq \bigcup \boldsymbol{X}$. Since A was an arbitrary element of $\boldsymbol{X}$, we have shown that for all $A \in \boldsymbol{X}$, we have $A \subseteq \bigcup \boldsymbol{X}$.

63. For all $A \in \boldsymbol{X}$, $\bigcap \boldsymbol{X} \subseteq A$.

Let $A \in \boldsymbol{X}$ and let $x \in \bigcap \boldsymbol{X}$. Then for every $B \in \boldsymbol{X}$, we have $x \in B$. In particular, $x \in A$ (because $A \in \boldsymbol{X}$). Since x was an arbitrary element of $\bigcap \boldsymbol{X}$, we have shown that $\bigcap \boldsymbol{X} \subseteq A$. Since A was an arbitrary element of $\boldsymbol{X}$, we have shown that for all $A \in \boldsymbol{X}$, we have $\bigcap \boldsymbol{X} \subseteq A$.

Let A be a set and let $\boldsymbol{X}$ be a nonempty set of sets. Verify each of the following:

64. $A \cap \bigcup \boldsymbol{X} = \bigcup\{A \cap B \mid B \in \boldsymbol{X}\}$.

$x \in A \cap \bigcup \boldsymbol{X} \Leftrightarrow x \in A$ and $x \in \bigcup \boldsymbol{X} \Leftrightarrow x \in A$ and there is a $B \in \boldsymbol{X}$ with $x \in B \Leftrightarrow x \in A \cap B$ for some $B \in \boldsymbol{X} \Leftrightarrow x \in \bigcup\{A \cap B \mid B \in \boldsymbol{X}\}$.

65. $A \cup \bigcap \boldsymbol{X} = \bigcap\{A \cup B \mid B \in \boldsymbol{X}\}$.

$x \in A \cup \bigcap \boldsymbol{X} \Leftrightarrow x \in A$ or $x \in \bigcap \boldsymbol{X} \Leftrightarrow x \in A$ or $x \in B$ for every $B \in \boldsymbol{X} \Leftrightarrow x \in A \cup B$ for every $B \in \boldsymbol{X} \Leftrightarrow$ $x \in \bigcap\{A \cup B \mid B \in \boldsymbol{X}\}$.

66. $A \setminus \bigcup \boldsymbol{X} = \bigcap\{A \setminus B \mid B \in \boldsymbol{X}\}$.

$x \in A \setminus \bigcup \boldsymbol{X} \Leftrightarrow x \in A$ and $x \notin \bigcup \boldsymbol{X} \Leftrightarrow x \in A$ and $x \notin B$ for every $B \in \boldsymbol{X} \Leftrightarrow x \in A \setminus B$ for every $B \in \boldsymbol{X} \Leftrightarrow x \in \bigcap\{A \setminus B \mid B \in \boldsymbol{X}\}$.

67. $A \setminus \bigcap \boldsymbol{X} = \bigcup\{A \setminus B \mid B \in \boldsymbol{X}\}$.

$x \in A \setminus \bigcap \boldsymbol{X} \Leftrightarrow x \in A$ and $x \notin \bigcap \boldsymbol{X} \Leftrightarrow x \in A$ and $x \notin B$ for some $B \in \boldsymbol{X} \Leftrightarrow x \in A \setminus B$ for some $B \in \boldsymbol{X} \Leftrightarrow x \in \bigcup\{A \setminus B \mid B \in \boldsymbol{X}\}$.

Find the natural number that each of the following expressions is equal to or explain why the given expression is not a natural number.

68. $\bigcup\{2k \mid 0 \leq k \leq 100\}$

$\bigcup\{2k \mid 0 \leq k \leq 100\} = 0 \cup 2 \cup 4 \cup \cdots \cup 200 = 0 \cup \{0,1\} \cup \{0,1,2,3\} \cup \cdots \cup \{0,1,2,\dots,199\}$
$= \{0,1,2,\dots,199\} = \mathbf{200}$.

69. $\bigcup\{\{k\} \mid 0 \leq k \leq 100\}$

$\bigcup\{\{k\} \mid 0 \leq k \leq 100\} = \{0\} \cup \{1\} \cup \{2\} \cup \cdots \cup \{100\} = \{0,1,2,\dots,100\} = \mathbf{101}$.

70. $\bigcup\{\{2k\} \mid 0 \leq k \leq 100\}$

$\bigcup\{\{2k\} \mid 0 \leq k \leq 100\} = \{0\} \cup \{2\} \cup \{4\} \cup \cdots \cup \{200\} = \{0,2,4,\dots,200\}$. This is **not** a natural number.

71. $202 \setminus \bigcup\{2k \mid 0 \leq k \leq 100\}$

By Problem 68, $\bigcup\{2k \mid 0 \leq k \leq 100\} = 200$. Therefore, $202 \setminus \bigcup\{2k \mid 0 \leq k \leq 100\} = 202 \setminus 200$ $= \{0, 1, 2, \dots, 201\} \setminus \{0, 1, 2, \dots, 199\} = \{200, 201\}$. This is **not** a natural number.

72. $(202 \setminus \bigcup\{2k \mid 0 \leq k \leq 100\}) \cup 201$

By Problem 71, we have $202 \setminus \bigcup\{2k \mid 0 \leq k \leq 100\} = \{200, 201\}$. Therefore, we see that $(202 \setminus \bigcup\{2k \mid 0 \leq k \leq 100\}) \cup 201 = \{200, 201\} \cup \{0, 1, 2, \dots, 200\} = \{0, 1, 2, \dots, 201\} = \mathbf{202}$.

Problem Set 3

LEVEL 1

Determine if each of the following is true or false.

1. $(4,5) = (5,4)$

False

2. $\{4,5\} = \{5,4\}$

True

3. $(4,5) = \{\{4\},\{4,5\}\}$

True

4. $(4,4,5) = (4,5,4)$

False

5. $\{4,4,5\} = \{4,5\}$

True

6. $(4,4,5) = (4,5)$

False

Use the roster method to describe each of the following sets.

7. $\{u\} \times \{v\} \times \{w\} \times \{z\} \times \{t\}$

$\{(\boldsymbol{u}, \boldsymbol{v}, \boldsymbol{w}, \boldsymbol{z}, \boldsymbol{t})\}$

8. $\{u, v, w\} \times \{z, t\}$

$\{(\boldsymbol{u}, \boldsymbol{z}), (\boldsymbol{u}, \boldsymbol{t}), (\boldsymbol{v}, \boldsymbol{z}), (\boldsymbol{v}, \boldsymbol{t}), (\boldsymbol{w}, \boldsymbol{z}), (\boldsymbol{w}, \boldsymbol{t})\}$

9. $\{z, t\}^2$

$\{z, t\}^2 = \{z, t\} \times \{z, t\} = \{(\boldsymbol{z}, \boldsymbol{z}), (\boldsymbol{z}, \boldsymbol{t}), (\boldsymbol{t}, \boldsymbol{z}), (\boldsymbol{t}, \boldsymbol{t})\}$

10. $\{z, t\}^3$

$\{z, t\}^3 = \{z, t\} \times \{z, t\} \times \{z, t\} = \{(\boldsymbol{z}, \boldsymbol{z}, \boldsymbol{z}), (\boldsymbol{z}, \boldsymbol{z}, \boldsymbol{t}), (\boldsymbol{z}, \boldsymbol{t}, \boldsymbol{z}), (\boldsymbol{z}, \boldsymbol{t}, \boldsymbol{t}), (\boldsymbol{t}, \boldsymbol{z}, \boldsymbol{z}), (\boldsymbol{t}, \boldsymbol{z}, \boldsymbol{t}), (\boldsymbol{t}, \boldsymbol{t}, \boldsymbol{z}), (\boldsymbol{t}, \boldsymbol{t}, \boldsymbol{t})\}$

11. $\{z, t\}^4$

$$\{z,t\}^4 = \{z,t\} \times \{z,t\} \times \{z,t\} \times \{z,t\}$$
$$= \{(\boldsymbol{z},\boldsymbol{z},\boldsymbol{z},\boldsymbol{z}), (\boldsymbol{z},\boldsymbol{z},\boldsymbol{z},\boldsymbol{t}), (\boldsymbol{z},\boldsymbol{z},\boldsymbol{t},\boldsymbol{z}), (\boldsymbol{z},\boldsymbol{z},\boldsymbol{t},\boldsymbol{t}), (\boldsymbol{z},\boldsymbol{t},\boldsymbol{z},\boldsymbol{z}), (\boldsymbol{z},\boldsymbol{t},\boldsymbol{z},\boldsymbol{t}), (\boldsymbol{z},\boldsymbol{t},\boldsymbol{t},\boldsymbol{z}), (\boldsymbol{z},\boldsymbol{t},\boldsymbol{t},\boldsymbol{t}),$$
$$(\boldsymbol{t},\boldsymbol{z},\boldsymbol{z},\boldsymbol{z}), (\boldsymbol{t},\boldsymbol{z},\boldsymbol{z},\boldsymbol{t}), (\boldsymbol{t},\boldsymbol{z},\boldsymbol{t},\boldsymbol{z}), (\boldsymbol{t},\boldsymbol{z},\boldsymbol{t},\boldsymbol{t}), (\boldsymbol{t},\boldsymbol{t},\boldsymbol{z},\boldsymbol{z}), (\boldsymbol{t},\boldsymbol{t},\boldsymbol{z},\boldsymbol{t}), (\boldsymbol{t},\boldsymbol{t},\boldsymbol{t},\boldsymbol{z}), (\boldsymbol{t},\boldsymbol{t},\boldsymbol{t},\boldsymbol{t})\},$$

12. $\{u,v,w\}^2$

$\{u,v,w\}^2 = \{u,v,w\} \times \{u,v,w\} = \{(\boldsymbol{u},\boldsymbol{u}), (\boldsymbol{u},\boldsymbol{v}), (\boldsymbol{u},\boldsymbol{w}), (\boldsymbol{v},\boldsymbol{u}), (\boldsymbol{v},\boldsymbol{v}), (\boldsymbol{v},\boldsymbol{w}), (\boldsymbol{w},\boldsymbol{u}), (\boldsymbol{w},\boldsymbol{v}), (\boldsymbol{w},\boldsymbol{w})\}$

13. $\{u,v,w\}^3$

$$\{u,v,w\}^3 = \{u,v,w\} \times \{u,v,w\} \times \{u,v,w\}$$
$$= \{(\boldsymbol{u},\boldsymbol{u},\boldsymbol{u}), (\boldsymbol{u},\boldsymbol{u},\boldsymbol{v}), (\boldsymbol{u},\boldsymbol{u},\boldsymbol{w}), (\boldsymbol{u},\boldsymbol{v},\boldsymbol{u}), (\boldsymbol{u},\boldsymbol{v},\boldsymbol{v}), (\boldsymbol{u},\boldsymbol{v},\boldsymbol{w}), (\boldsymbol{u},\boldsymbol{w},\boldsymbol{u}), (\boldsymbol{u},\boldsymbol{w},\boldsymbol{v}), (\boldsymbol{u},\boldsymbol{w},\boldsymbol{w}),$$
$$(\boldsymbol{v},\boldsymbol{u},\boldsymbol{u}), (\boldsymbol{v},\boldsymbol{u},\boldsymbol{v}), (\boldsymbol{v},\boldsymbol{u},\boldsymbol{w}), (\boldsymbol{v},\boldsymbol{v},\boldsymbol{u}), (\boldsymbol{v},\boldsymbol{v},\boldsymbol{v}), (\boldsymbol{v},\boldsymbol{v},\boldsymbol{w}), (\boldsymbol{v},\boldsymbol{w},\boldsymbol{u}), (\boldsymbol{v},\boldsymbol{w},\boldsymbol{v}), (\boldsymbol{v},\boldsymbol{w},\boldsymbol{w}),$$
$$(\boldsymbol{w},\boldsymbol{u},\boldsymbol{u}), (\boldsymbol{w},\boldsymbol{u},\boldsymbol{v}), (\boldsymbol{w},\boldsymbol{u},\boldsymbol{w}), (\boldsymbol{w},\boldsymbol{v},\boldsymbol{u}), (\boldsymbol{w},\boldsymbol{v},\boldsymbol{v}), (\boldsymbol{w},\boldsymbol{v},\boldsymbol{w}), (\boldsymbol{w},\boldsymbol{w},\boldsymbol{u}), (\boldsymbol{w},\boldsymbol{w},\boldsymbol{v}), (\boldsymbol{w},\boldsymbol{w},\boldsymbol{w})\}$$

14. $\emptyset^2$

$\emptyset^2 = \emptyset \times \emptyset = \emptyset$

15. $\emptyset^3$

$\emptyset^3 = \emptyset \times \emptyset \times \emptyset = \emptyset$

16. $\left(\mathcal{P}(\{\emptyset\})\right)^2$

Since $\mathcal{P}(\{\emptyset\}) = \{\emptyset, \{\emptyset\}\}$, we have

$$\mathcal{P}(\{\emptyset\})^2 = \mathcal{P}(\{\emptyset\}) \times \mathcal{P}(\{\emptyset\}) = \{(\emptyset,\emptyset), (\emptyset,\{\emptyset\}), (\{\emptyset\},\emptyset), (\{\emptyset\},\{\emptyset\})\}.$$

Let S be the set of people on a checkout line at a supermarket and define a relation $\prec$ on X by $x \prec y$ if x will check out before y.

17. Is $\prec$ reflexive on S?

No

18. Is $\prec$ symmetric on S?

No

19. Is $\prec$ transitive on S?

Yes

20. Is $\prec$ antireflexive on S?

Yes

21. Is $\prec$ antisymmetric on S?

Yes (vacuously)

22. Is $\prec$ trichotomous on S?

Yes

23. Does $\prec$ satisfy the comparability condition on S?

No

24. Is $(S, \prec)$ a poset?

No

25. Is $(S, \prec)$ a strict poset?

Yes

26. Is $(S, \prec)$ an ordered set?

Yes

Determine if each of the following is an ordered set.

27. $(\mathbb{Z}, >)$

Yes

28. $(\mathbb{Q}, <)$

Yes

29. $(\mathbb{Q}, \leq)$

No

30. $(\mathbb{R}, >)$

Yes

LEVEL 2

Find the domain, range, and field of each of the following relations:

31. $R = \{(a, b), (c, d), (e, f), (f, a)\}$

$\textbf{dom } \boldsymbol{R} = \{\boldsymbol{a}, \boldsymbol{c}, \boldsymbol{e}, \boldsymbol{f}\}$; $\textbf{ran } \boldsymbol{R} = \{\boldsymbol{a}, \boldsymbol{b}, \boldsymbol{d}, \boldsymbol{f}\}$; $\textbf{field } \boldsymbol{R} = \{\boldsymbol{a}, \boldsymbol{b}, \boldsymbol{c}, \boldsymbol{d}, \boldsymbol{e}, \boldsymbol{f}\}$

32. $S = \{(2k, 2t+1) \mid k, t \in \mathbb{Z}\}$

$\mathbf{dom}\ \boldsymbol{S} = \mathbf{2}\mathbb{Z} = \mathbb{E}$; $\mathbf{ran}\ \boldsymbol{S} = \{\mathbf{2}\boldsymbol{t} + \mathbf{1} \mid \boldsymbol{t} \in \mathbb{Z}\} = \mathbf{2}\mathbb{Z} + \mathbf{1} = \mathbb{O}$; $\mathbf{field}\ \boldsymbol{S} = \mathbb{Z}$

33. $T = \left\{(a, b) \,\middle|\, \frac{a}{b} \in \mathbb{Q}\right\}$

$\mathbf{dom}\ \boldsymbol{T} = \mathbb{Z}$; $\mathbf{ran}\ \boldsymbol{T} = \mathbb{Z}^*$; $\mathbf{field}\ \boldsymbol{T} = \mathbb{Z}$

34. $U = \mathbb{R} \times \mathbb{Z}^+$

$\mathbf{dom}\ \boldsymbol{T} = \mathbb{R}$; $\mathbf{ran}\ \boldsymbol{T} = \mathbb{Z}^+$; $\mathbf{field}\ \boldsymbol{T} = \mathbb{R}$

Determine a and b so that each of the following equations is true, or state that the equation has no solution.

35. $\{a, b\} = \{8\}$

$\boldsymbol{a} = \mathbf{8}, \boldsymbol{b} = \mathbf{8}$

36. $\{a, b\} = \{7, 9\}$

$\boldsymbol{a} = \mathbf{7}, \boldsymbol{b} = \mathbf{9}$ or $\boldsymbol{a} = \mathbf{9}, \boldsymbol{b} = \mathbf{7}$

37. $(a, b) = (7, 9)$

$\boldsymbol{a} = \mathbf{7}, \boldsymbol{b} = \mathbf{9}$

38. $(a, b) = \{\{7\}, \{9, 7\}\}$

$\boldsymbol{a} = \mathbf{7}, \boldsymbol{b} = \mathbf{9}$ (note that $\{\{7\}, \{9, 7\}\} = \{\{7\}, \{7, 9\}\}$)

39. $(a, b) = \{5\}$

No solution.

40. $(a, b) = \{\{5\}\}$

$\boldsymbol{a} = \mathbf{5}, \boldsymbol{b} = \mathbf{5}$

41. $(a, a, b) = (1, b, a)$

$\boldsymbol{a} = \mathbf{1}, \boldsymbol{b} = \mathbf{1}$

42. $(a, b, c, d) = (1, c, 2, a)$

$\boldsymbol{a} = \mathbf{1}, \boldsymbol{b} = \mathbf{2}, \boldsymbol{c} = \mathbf{2}, \boldsymbol{d} = \mathbf{1}$

Suppose that $|A| = 5$, $|B| = 10$, $|C| = 50$, and $|D| = 100$. Compute each of the following:

43. $|A \times B|$

$|A \times B| = |A| \cdot |B| = 5 \cdot 10 = \mathbf{50}$

44. $|A \times B \times C \times D|$

$|A \times B \times C \times D| = |A| \cdot |B| \cdot |C| \cdot |D| = 5 \cdot 10 \cdot 50 \cdot 100 = \mathbf{250,000}$

45. $|C^2|$

$|C^2| = |C|^2 = 50^2 = \mathbf{2500}$

46. $\mathcal{P}(C \times D)$

$|\mathcal{P}(C \times D)| = 2^{|C \times D|} = 2^{|C| \times |D|} = 2^{50 \cdot 100} = \mathbf{2^{5000}}$

47. $\{R \mid R \text{ is a relation on } B\}$

$|\mathcal{P}(B \times B)| = 2^{|B \times B|} = 2^{|B| \times |B|} = 2^{10 \cdot 10} = \mathbf{2^{100}}$

48. $\{R \mid R \text{ is a relation on } B \times C\}$

$|\mathcal{P}((B \times C) \times (B \times C))| = 2^{|(B \times C) \times (B \times C)|} = 2^{|B \times C| \times |B \times C|} = 2^{500 \cdot 500} = \mathbf{2^{250,000}}$

LEVEL 3

Write each of the following in its unabbreviated form:

49. (x, x, z)

$(x, x, z) = ((x, x), z) = \{\{(x, x)\}, \{(x, x), z\}\} = \boldsymbol{\{\{\{\{x\}\}\}, \{\{\{x\}\}, z\}\}}$

50. (x, z, x)

$(x, z, x) = ((x, z), x) = \{\{(x, z)\}, \{(x, z), x\}\} = \boldsymbol{\{\{\{\{x\}, \{x, z\}\}\}, \{\{\{x\}, \{x, z\}\}, x\}\}}$

51. (x, x, x, x)

$(x, x, x, x) = ((x, x, x), x) = \boldsymbol{\{\{\{\{\{\{x\}\}\}, \{\{\{x\}\}, x\}\}\}, \{\{\{\{\{x\}\}\}, \{\{\{x\}\}, x\}\}, x\}\}}$

52. (x, y, z, w)

$$(x,y,z,w) = \big((x,y,z),w\big) = \{\{(x,y,z)\},\{(x,y,z),w\}\}$$

$$= \left\{\left\{\left\{\left\{\{\{x\},\{x,y\}\}\right\},\left\{\{\{x\},\{x,y\}\},z\right\}\right\}\right\},\left\{\left\{\left\{\{\{x\},\{x,y\}\}\right\},\left\{\{\{x\},\{x,y\}\},z\right\}\right\},w\right\}\right\}$$

Let $A = \{a,b\}$ with $a \neq b$, consider the strict posets $(\mathbb{Z},<)$ and $(\mathcal{P}(A),\subset)$, and let $<_D$ be the dictionary order on $\mathbb{Z} \times \mathcal{P}(A)$.

53. Is $<_D$ reflexive on $\mathbb{Z} \times \mathcal{P}(A)$?

No

54. Is $<_D$ symmetric on $\mathbb{Z} \times \mathcal{P}(A)$?

No

55. Is $<_D$ transitive on $\mathbb{Z} \times \mathcal{P}(A)$?

Yes

56. Is $<_D$ antireflexive on $\mathbb{Z} \times \mathcal{P}(A)$?

Yes

57. Is $<_D$ antisymmetric on $\mathbb{Z} \times \mathcal{P}(A)$?

Yes

58. Is $<_D$ trichotomous on $\mathbb{Z} \times \mathcal{P}(A)$?

No

59. Does $<_D$ satisfy the comparability condition on $\mathbb{Z} \times \mathcal{P}(A)$?

No

60. Is $(\mathbb{Z} \times \mathcal{P}(A), <_D)$ a poset?

No

61. Is $(\mathbb{Z} \times \mathcal{P}(A), <_D)$ a strict poset?

Yes

62. Is $(\mathbb{Z} \times \mathcal{P}(A), <_D)$ an ordered set?

No

Determine if each of the following is true or false.

63. $\{\emptyset, \{\emptyset\}\} \in 3$

True because $\{\emptyset, \{\emptyset\}\} = \{0, 1\} = 2$ and $2 \in 3$ (or equivalently, $\{\emptyset, \{\emptyset\}\} \in \{\emptyset, \{\emptyset\}, \{\emptyset, \{\emptyset\}\}\}$).

64. $3 \in \{\emptyset, \{\emptyset\}, \{\emptyset, \{\emptyset\}\}\}$

False because $\{\emptyset, \{\emptyset\}, \{\emptyset, \{\emptyset\}\}\} = \{0, 1, 2\} = 3$ and $3 \notin 3$.

65. $5 \in 8$

True because $(\mathbb{N}, <)$ is an ordered set and $5 \in 8$ is equivalent to $5 < 8$.

66. $5 <_{\mathbb{N}} \{0, 1, 2, 3, 4, 5, 6, 7\}$

True because $\{0, 1, 2, 3, 4, 5, 6, 7\} = 8$ and $5 \in 8$.

LEVEL 4

Let A, B, C, and D be sets. Determine if each of the following statements is true or false. If true, explain why. If false, provide a counterexample.

67. $(A \times B) \cap (C \times D) = (A \cap C) \times (B \cap D)$.

True.

$(x, y) \in (A \times B) \cap (C \times D)$ if and only if $(x, y) \in A \times B$ and $(x, y) \in C \times D$ if and only if $x \in A, y \in B$, $x \in C$, and $y \in D$ if and only if $x \in A \cap C$ and $y \in B \cap D$ if and only if $(x, y) \in (A \cap C) \times (B \cap D)$. Therefore, $(A \times B) \cap (C \times D) = (A \cap C) \times (B \cap D)$.

68. $(A \times B) \cup (C \times D) = (A \cup C) \times (B \cup D)$.

False.

Let $A = \{0\}, B = \{1\}, C = \{2\}, D = \{3\}$.

Then $A \times B = \{(0, 1)\}$, $C \times D = \{(2, 3)\}$, and so, $(A \times B) \cup (C \times D) = \{(0, 1), (2, 3)\}$.

Also, $A \cup C = \{0, 2\}$, $B \cup D = \{1, 3\}$, and so, $(A \cup C) \times (B \cup D) = \{(0, 1), (0, 3), (2, 1), (2, 3)\}$.

Since $(2, 1) \in (A \cup C) \times (B \cup D)$, but $(2, 1) \notin (A \times B) \cup (C \times D)$, we see that

$$(A \times B) \cup (C \times D) \neq (A \cup C) \times (B \cup D).$$

69. If $A \subseteq B$ and $C \subseteq D$, then $A \times C \subseteq B \times D$.

True.

Let $(x, y) \in A \times C$. Then $x \in A$ and $y \in C$. Since $x \in A$ and $A \subseteq B$, $x \in B$. Since $y \in C$ and $C \subseteq D$, $y \in D$. Therefore, $(x, y) \in B \times D$. Since $(x, y) \in A \times C$ was arbitrary, $A \times C \subseteq B \times D$.

Using the definition $(x, y) = \{\{x\}, \{x, y\}\}$, verify that each of the following is true.

70. If $x = z$ and $y = w$, then $(x, y) = (z, w)$.

Since $x = z$, $\{x\} = \{z\}$. Since $x = z$ and $y = w$, $\{x, y\} = \{z, w\}$. So, we have

$$(x, y) = \{\{x\}, \{x, y\}\} = \{\{z\}, \{z, w\}\} = (z, w).$$

71. If $x \neq y$ and $\{\{x\}, \{x, y\}\} = \{\{z\}, \{z, w\}\}$, then $x = z$.

Since $x \neq y$, $|\{x, y\}| = 2$. Since $|\{z\}| = 1$, we see that $\{x, y\} \neq \{z\}$. So, $\{x, y\} = \{z, w\}$ and it follows that $\{x\} = \{z\}$. So, $x = z$.

72. If $x \neq y$ and $\{\{x\}, \{x, y\}\} = \{\{z\}, \{z, w\}\}$, then $y = w$.

By Problem 71, $x = z$ and $\{x, y\} = \{z, w\}$. So, $y = w$.

73. If $x \neq y$ and $(x, y) = (z, w)$, then $x = z$ and $y = w$.

Since $(x, y) = \{\{x\}, \{x, y\}\}$ and $(z, w) = \{\{z\}, \{z, w\}\}$, we have $\{\{x\}, \{x, y\}\} = \{\{z\}, \{z, w\}\}$. By Problem 71, $x = z$. By Problem 72, $y = w$.

74. $\{\{x\}, \{x, x\}\} = \{\{x\}\}$.

$\{\{x\}, \{x, x\}\} = \{\{x\}, \{x\}\} = \{\{x\}\}$.

75. If $(x, x) = (z, w)$, then $x = z = w$.

By Problem 74, $(x, x) = \{\{x\}, \{x, x\}\} = \{\{x\}\}$. So, $\{\{z\}, \{z, w\}\} = (z, w) = (x, x) = \{\{x\}\}$. It follows that $\{z\} = \{x\}$ and $\{z, w\} = \{x\}$. Since $\{z, w\} = \{x\}$, $z = x$ and $w = x$.

76. If $(x, y) = (z, w)$, then $x = z$ and $y = w$.

If $x \neq y$, then by Problem 73, $x = z$ and $y = w$. If $x = y$, then by Problem 75, $x = z = w$. In particular, $x = z$ and $y = w$.

Verify that each of the following is true.

77. If $x = u$, $y = v$, and $z = w$, then $(x, y, z) = (u, v, w)$.

If $x = u$, $y = v$, and $z = w$, then $(x, y) = (u, v)$ and $z = w$. So, $((x, y), z) = ((u, v), w)$. Therefore, $(x, y, z) = ((x, y), z) = ((u, v), w) = (u, v, w)$.

78. If $(x, y, z) = (u, v, w)$, then $x = u$, $y = v$, and $z = w$.

If $(x, y, z) = (u, v, w)$, then $\big((x, y), z\big) = \big((u, v), w\big)$. By Problem 76, $(x, y) = (u, v)$ and $z = w$. Again by Problem 76, $x = u$ and $y = v$.

79. If $x_1 = y_1$, $x_2 = y_2$, $x_3 = y_3$, and $x_4 = y_4$, then $(x_1, x_2, x_3, x_4) = (y_1, y_2, y_3, y_4)$.

If $x_1 = y_1$, $x_2 = y_2$, $x_3 = y_3$, and $x_4 = y_4$, then by Problem 77, $(x_1, x_2, x_3) = (y_1, y_2, y_3)$. So, we have $(x_1, x_2, x_3, x_4) = \big((x_1, x_2, x_3), x_4\big) = \big((y_1, y_2, y_3), y_4\big) = (y_1, y_2, y_3, y_4)$.

80. If $(x_1, x_2, x_3, x_4) = (y_1, y_2, y_3, y_4)$, then $x_1 = y_1$, $x_2 = y_2$, $x_3 = y_3$, and $x_4 = y_4$.

If $(x_1, x_2, x_3, x_4) = (y_1, y_2, y_3, y_4)$, then $\big((x_1, x_2, x_3), x_4\big) = \big((y_1, y_2, y_3), y_4\big)$. By Problem 76, we have $(x_1, x_2, x_3) = (y_1, y_2, y_3)$ and $x_4 = y_4$. By Problem 78, $x_1 = y_1$, $x_2 = y_2$, and $x_3 = y_3$.

LEVEL 5

Let R be a relation on a set A. Determine if each of the following statements is true or false. If true, explain why. If false, provide a counterexample.

81. If R is symmetric and transitive on A, then R is reflexive on A.

This is **false**. Let $A = \{0, 1\}$ and $R = \{(0, 0)\}$. Then R is symmetric and transitive, but not reflexive (because $(1, 1) \notin R$).

82. If R is antisymmetric on A, then R is not symmetric on A.

This is **false**. $\emptyset$ is both symmetric and antisymmetric on any set A.

For $a, b \in \mathbb{N}$, we will say that a divides b, written $a|b$, if there is a natural number k such that $b = ak$. Notice that $|$ is a binary relation on $\mathbb{N}$.

83. Is $|$ reflexive on $\mathbb{N}$?

Yes. If $a \in \mathbb{N}$ then $a = 1a$, so that $a|a$.

84. Is $|$ symmetric on $\mathbb{N}$?

No. For example, $2|6$ (because $6 = 2 \cdot 3$), but $2 \neq 6 \cdot k$ for any natural number k.

85. Is $|$ transitive on $\mathbb{N}$?

Yes. If $a|b$ and $b|c$, then there are natural numbers j and k such that $b = ja$ and $c = kb$. Then $c = kb = k(ja) = (kj)a$. Since the product of two natural numbers is a natural number, $kj \in \mathbb{N}$. So, $a|c$.

86. Is $|$ antireflexive on $\mathbb{N}$?

No. For example, $2|2$ (because $2 = 2 \cdot 1$).

87. Is $|$ antisymmetric on $\mathbb{N}$?

Yes. If $a|b$ and $b|a$, then there are natural numbers j and k such that $b = ja$ and $a = kb$. If $a = 0$, then $b = j \cdot 0 = 0$, and so, $a = b$. Suppose $a \neq 0$. We have $a = k(ja) = (kj)a$. Therefore, $(kj - 1)a = (kj)a - 1a = 0$. So, $kj - 1 = 0$, and therefore, $kj = 1$. So, $k = j = 1$. Thus, $b = ja = 1a = a$.

88. Is $|$ trichotomous on $\mathbb{N}$?

No. For example, 2 and 3 are distinct and do not divide each other.

89. Does $|$ satisfy the comparability condition on $\mathbb{N}$?

No. For example, 2 and 3 do not divide each other.

90. Is $(\mathbb{N}, |)$ a poset?

Yes because $|$ is reflexive (by Problem 83), antisymmetric (by Problem 87), and transitive (by Problem 85).

91. Is $(\mathbb{N}, |)$ a strict poset?

No because $|$ is **not** antireflexive (by Problem 86).

92. Is $(\mathbb{N}, |)$ an ordered set?

No because $|$ is **not** trichotomous (by Problem 88).

Problem Set 4

Full solutions to these problems are available for free download here:

www.SATPrepGet800.com/STKZ3D

LEVEL 1

Let $X = \{a, b, c\}$. Determine if each of the following is an equivalence relation on X.

1. $\{(a,a),(b,b),(c,c)\}$

Yes. This is the equality relation on X.

2. $\{(a,a),(a,b),(a,c)\}$

No. This relation is not reflexive (for example, it does not include (b,b)) or symmetric (for example, it includes (a,b), but not (b,a)).

3. $\{(a,a),(b,a),(c,a)\}$

No. This relation is not reflexive (for example, it does not include (b,b)) or symmetric (for example, it includes (b,a), but not (a,b)).

4. $\{(a,a),(a,b),(a,c),(b,a),(b,b),(b,c),(c,a),(c,b),(c,c)\}$

Yes. This is the trivial equivalence relation on X.

5. $\{(a,a),(a,b),(a,c),(b,b),(b,c),(c,c)\}$

No. This relation is not symmetric (for example, it includes (a,b), but not (b,a)).

6. $\{(a,a),(a,b),(b,b),(b,a),(c,c)\}$

Yes. This relation is reflexive, symmetric, and transitive.

7. $\{(a,a),(a,b),(b,a),(b,b),(b,c),(c,b),(c,c)\}$

No. This relation is not transitive (for example, it includes (a,b) and (b,c), but not (a,c)).

8. $\{(a,a),(b,b),(b,c),(c,c),(c,b)\}$

Yes. This relation is reflexive, symmetric, and transitive.

Let $X = \{0, 1, 2, 3, 4\}$. For each of the following, describe a relation R on X with the given properties:

9. R is reflexive, but not symmetric.

$\boldsymbol{R = \{(0,0),(0,1),(1,1),(2,2),(3,3),(4,4)\}}$. Notice that $(0,1) \in R$, but $(1,0) \notin R$.

10. R is reflexive, but not transitive.

$\boldsymbol{R} = \{(\mathbf{0},\mathbf{0}),(\mathbf{0},\mathbf{1}),(\mathbf{1},\mathbf{1}),(\mathbf{1},\mathbf{2}),(\mathbf{2},\mathbf{2}),(\mathbf{3},\mathbf{3}),(\mathbf{4},\mathbf{4})\}$. Notice that $(0,1),(1,2) \in R$, but $(0,2) \notin R$.

11. R is symmetric, but not reflexive.

$\boldsymbol{R} = \{(\mathbf{0},\mathbf{1}),(\mathbf{1},\mathbf{0})\}$. Notice that $(0,0) \notin R$.

12. R is symmetric, but not transitive.

$\boldsymbol{R} = \{(\mathbf{0},\mathbf{1}),(\mathbf{1},\mathbf{0}),(\mathbf{1},\mathbf{2}),(\mathbf{2},\mathbf{1})\}$. Notice that $(0,1),(1,2) \in R$, but $(0,2) \notin R$.

13. R is transitive, but not reflexive.

$\boldsymbol{R} = \{(\mathbf{0},\mathbf{1}),(\mathbf{0},\mathbf{2}),(\mathbf{1},\mathbf{2})\}$. Notice that $(0,0) \notin R$.

14. R is transitive, but not symmetric.

$\boldsymbol{R} = \{(\mathbf{0},\mathbf{1}),(\mathbf{0},\mathbf{2}),(\mathbf{1},\mathbf{2})\}$. Notice that $(0,1) \in R$, but $(1,0) \notin R$.

Find all partitions of each of the following sets:

15. $\{x\}$

There is only one partition of $\{x\}$, namely $\{\{x\}\}$.

16. $\{x,y\}$

The partitions of $\{x,y\}$ are $\{\{\boldsymbol{x}\},\{\boldsymbol{y}\}\}$ and $\{\{\boldsymbol{x},\boldsymbol{y}\}\}$.

17. $\{x,y,z\}$

The partitions of $\{x,y,z\}$ are $\{\{\boldsymbol{x}\},\{\boldsymbol{y}\},\{\boldsymbol{z}\}\}$, $\{\{\boldsymbol{x}\},\{\boldsymbol{y},\boldsymbol{z}\}\}$, $\{\{\boldsymbol{y}\},\{\boldsymbol{x},\boldsymbol{z}\}\}$, $\{\{\boldsymbol{z}\},\{\boldsymbol{x},\boldsymbol{y}\}\}$, and $\{\{\boldsymbol{x},\boldsymbol{y},\boldsymbol{z}\}\}$.

18. $\{x,y,z,w\}$

The partitions of $\{x,y,z,w\}$ are $\{\{\boldsymbol{x}\},\{\boldsymbol{y}\},\{\boldsymbol{z}\},\{\boldsymbol{w}\}\}$, $\{\{\boldsymbol{x}\},\{\boldsymbol{y}\},\{\boldsymbol{z},\boldsymbol{w}\}\}$, $\{\{\boldsymbol{x}\},\{\boldsymbol{z}\},\{\boldsymbol{y},\boldsymbol{w}\}\}$, $\{\{\boldsymbol{x}\},\{\boldsymbol{w}\},\{\boldsymbol{y},\boldsymbol{z}\}\}$, $\{\{\boldsymbol{y}\},\{\boldsymbol{z}\},\{\boldsymbol{x},\boldsymbol{w}\}\}$, $\{\{\boldsymbol{y}\},\{\boldsymbol{w}\},\{\boldsymbol{x},\boldsymbol{z}\}\}$, $\{\{\boldsymbol{z}\},\{\boldsymbol{w}\},\{\boldsymbol{x},\boldsymbol{y}\}\}$, $\{\{\boldsymbol{x},\boldsymbol{y}\},\{\boldsymbol{z},\boldsymbol{w}\}\}$, $\{\{\boldsymbol{x},\boldsymbol{z}\},\{\boldsymbol{y},\boldsymbol{w}\}\}$, $\{\{\boldsymbol{x},\boldsymbol{w}\},\{\boldsymbol{y},\boldsymbol{z}\}\}$, $\{\{\boldsymbol{x},\boldsymbol{y},\boldsymbol{z}\},\{\boldsymbol{w}\}\}$, $\{\{\boldsymbol{x},\boldsymbol{y},\boldsymbol{w}\},\{\boldsymbol{z}\}\}$, $\{\{\boldsymbol{x},\boldsymbol{z},\boldsymbol{w}\},\{\boldsymbol{y}\}\}$, $\{\{\boldsymbol{y},\boldsymbol{z},\boldsymbol{w}\},\{\boldsymbol{x}\}\}$, and $\{\{\boldsymbol{x},\boldsymbol{y},\boldsymbol{z},\boldsymbol{w}\}\}$.

LEVEL 2

For each of the following equivalence relations on $A = \{a,b,c,d\}$, find $[a]$, $[b]$, $[c]$, and $[d]$.

19. $\{(a,a),(b,b),(c,c),(d,d)\}$

$[\boldsymbol{a}] = \{\boldsymbol{a}\}$ $[\boldsymbol{b}] = \{\boldsymbol{b}\}$ $[\boldsymbol{c}] = \{\boldsymbol{c}\}$ $[\boldsymbol{d}] = \{\boldsymbol{d}\}$

20. $\{(a,a),(a,b),(b,a),(b,b),(c,c),(d,d)\}$

$[\boldsymbol{a}] = [\boldsymbol{b}] = \{\boldsymbol{a},\boldsymbol{b}\}$ $[\boldsymbol{c}] = \{\boldsymbol{c}\}$ $[\boldsymbol{d}] = \{\boldsymbol{d}\}$

21. $\{(a,a),(a,b),(a,c),(b,a),(b,b),(b,c),(c,a),(c,b),(c,c),(d,d)\}$

$[\boldsymbol{a}] = [\boldsymbol{b}] = [\boldsymbol{c}] = \{\boldsymbol{a},\boldsymbol{b},\boldsymbol{c}\}$ $[\boldsymbol{d}] = \{\boldsymbol{d}\}$

22. $\{(a,a),(a,c),(b,b),(b,d),(c,a),(c,c),(d,b),(d,d)\}$

$[\boldsymbol{a}] = [\boldsymbol{c}] = \{\boldsymbol{a},\boldsymbol{c}\}$ $[\boldsymbol{b}] = [\boldsymbol{d}] = \{\boldsymbol{b},\boldsymbol{d}\}$

Find an equivalence relation R on $A = \{0,1,2\}$ with the given property.

23. If S is any equivalence relation on A, then $R \subseteq S$.

$\boldsymbol{R} = \{(\boldsymbol{0},\boldsymbol{0}),(\boldsymbol{1},\boldsymbol{1}),(\boldsymbol{2},\boldsymbol{2})\}$

24. If S is any equivalence relation on A, then $S \subseteq R$.

$\boldsymbol{R} = \{(\boldsymbol{0},\boldsymbol{0}),(\boldsymbol{0},\boldsymbol{1}),(\boldsymbol{0},\boldsymbol{2}),(\boldsymbol{1},\boldsymbol{0}),(\boldsymbol{1},\boldsymbol{1}),(\boldsymbol{1},\boldsymbol{2}),(\boldsymbol{2},\boldsymbol{0}),(\boldsymbol{2},\boldsymbol{1}),(\boldsymbol{2},\boldsymbol{2})\}$

25. $|R| = 5$ and $0R1$.

$\boldsymbol{R} = \{(\boldsymbol{0},\boldsymbol{0}),(\boldsymbol{0},\boldsymbol{1}),(\boldsymbol{1},\boldsymbol{0}),(\boldsymbol{1},\boldsymbol{1}),(\boldsymbol{2},\boldsymbol{2})\}$

Determine if each of the following is a partition of $\{0,1,2,3,4,5,6,7,8\}$.

26. $\{\{0\},\{1\},\{2\},\{3\},\{4\},\{5\},\{6\},\{7\},\{8\}\}$

Yes

27. $\{\{0,1\},\{1,2,3\},\{3,4,5,6\},\{6,7,8\}\}$

No because for example, $\{0,1\} \cap \{1,2,3\} = \{1\} \neq \emptyset$.

28. $\{\{0,1,2,3,4,5,6,7,8\}\}$

Yes

29. $\{\{0,1,2,3\},\{5,6,7,8\}\}$

No because $\{0,1,2,3\} \cup \{5,6,7,8\} = \{1,2,3,5,6,7,8\} \neq \{1,2,3,4,5,6,7,8\}$.

30. $\{\{0,2,4,6,8\},\{1,3,5,7\}\}$

Yes

LEVEL 3

Let $X = \{0, 1, 2, 3, 4\}$. For each of the following, find the equivalence relation R on X of least cardinality containing the given elements.

31. $(0, 1) \in R$.

$\boldsymbol{R = \{(0,0), (0,1), (1,0), (1,1), (2,2), (3,3), (4,4)\}}$

32. $(0, 1), (2, 3) \in R$.

$\boldsymbol{R = \{(0,0), (0,1), (1,0), (1,1), (2,2), (2,3), (3,2), (3,3), (4,4)\}}$

33. $(0, 1), (1, 2), (2, 3) \in R$.

$\boldsymbol{R = \{(0,0), (0,1), (0,2), (0,3), (1,0), (1,1), (1,2), (1,3), (2,0), (2,1), (2,2), (2,3), (3,0), (3,1), (3,2), (3,3), (4,4)\}}$

34. $(0, 1), (2, 1), (3, 4), (4, 2) \in R$.

$\boldsymbol{R = \{(0,0), (0,1), (0,2), (0,3), (0,4), (1,0), (1,1), (1,2), (1,3), (1,4), (2,0), (2,1), (2,2), (2,3), (2,4), (3,0), (3,1), (3,2), (3,3), (3,4), (4,0), (4,1), (4,2), (4,3), (4,4)\}}$

Determine if each of the following is a partition of $\mathbb{Z}$.

35. $\{\mathbb{Z}\}$

Yes

36. $\{3\mathbb{Z}, 3\mathbb{Z} + 1\}$

No because $2 \notin 3\mathbb{Z} \cup (3\mathbb{Z} + 1)$.

37. $\{\mathbb{Z}^-, \mathbb{Z}^+\}$

No because $0 \notin \mathbb{Z}^- \cup \mathbb{Z}^+$.

38. $\{\mathbb{N}, \{-n \mid n \in \mathbb{N}\}\}$

No because $\mathbb{N} \cap \{-n \mid n \in \mathbb{N}\} = \{0\} \neq \emptyset$.

39. $\{5\mathbb{Z}, 5\mathbb{Z} + 1, 5\mathbb{Z} + 2, 5\mathbb{Z} + 3, 5\mathbb{Z} + 4\}$

Yes

Let $R = \{((a,b),(c,d)) \in (\mathbb{N} \times \mathbb{N})^2 \mid a + d = b + c\}$, so that $\mathbb{Z} = \{[(a,b)] \mid (a,b) \in \mathbb{N} \times \mathbb{N}\}$. Determine if each of the following is true or false.

40. $[(11,12)] = [(12,11)]$

False: $11 + 11 = 22$ and $12 + 12 = 24$.

41. $[(9,9)] = 0$

True

42. $[(0,1)] = 1$

False: $[(0,1)] = -1$

43. $[(9,7)] < [(7,9)]$

False: $9 + 9 = 18$, $7 + 7 = 14$, and $18 > 14$.

44. $[(0,2)] < [(3,5)]$

False: $0 + 5 = 2 + 3$

45. $[(4,4)] > [(3,4)]$

True: $4 + 4 > 4 + 3$

LEVEL 4

For $n, m \in \mathbb{Z}$, we will say that n divides m, written $n|m$, if there is an integer k such that $m = nk$. For $n \in \mathbb{Z}^+$, define $\equiv_n$ on $\mathbb{Z}$ by $\equiv_n = \{(a,b) \in \mathbb{Z}^2 \mid n|b - a\}$.

46. Show that $\equiv_n$ is reflexive on $\mathbb{Z}$.

Let $a \in \mathbb{Z}$. Then $a - a = 0 = n \cdot 0$. So, $n|a - a$. Therefore, $a \equiv_n a$, and so, $\equiv_n$ is reflexive.

47. Show that $\equiv_n$ is symmetric on $\mathbb{Z}$.

Let $a, b \in \mathbb{Z}$ and suppose that $a \equiv_n b$. Then $n|b - a$. So, there is $k \in \mathbb{Z}$ such that $b - a = nk$. Thus, $a - b = -(b - a) = -nk = n(-k)$. Since $k \in \mathbb{Z}$, $-k \in \mathbb{Z}$. So, $n|a - b$, and therefore, $b \equiv_n a$. So, $\equiv_n$ is symmetric.

48. Show that $\equiv_n$ is transitive on $\mathbb{Z}$.

Let $a, b, c \in \mathbb{Z}$ with $a \equiv_n b$ and $b \equiv_n c$. Then $n|b - a$ and $n|c - b$. So, there are $j, k \in \mathbb{Z}$ such that $b - a = nj$ and $c - b = nk$. So, $c - a = (c - b) + (b - a) = nk + nj = n(k + j)$. Since $j, k \in \mathbb{Z}$, $k + j \in \mathbb{Z}$. Therefore, $n|c - a$. So, $a \equiv_n c$. Thus, $\equiv_n$ is transitive.

49. Is $\equiv_n$ an equivalence relation on $\mathbb{Z}$?

Yes: Since $\equiv_n$ is reflexive, symmetric, and transitive, $\equiv_n$ is an equivalence relation on $\mathbb{Z}$

Let $\boldsymbol{X}$ be a set of equivalence relations on a set A. For each of the following statements, either verify that the statement is true or provide a counterexample showing that the statement is false.

50. $\bigcap \boldsymbol{X}$ is reflexive on A.

True: Let $x \in A$ and let $R \in \boldsymbol{X}$. Since R is reflexive, $(x, x) \in R$. Since $R \in \boldsymbol{X}$ was arbitrary, $\forall R \in \boldsymbol{X}\big((x, x) \in R\big)$. So, $(x, x) \in \bigcap \boldsymbol{X}$. Since $x \in A$ was arbitrary, $\bigcap \boldsymbol{X}$ is reflexive.

51. $\bigcup \boldsymbol{X}$ is reflexive on A.

True: Let $x \in A$ and let $R \in \boldsymbol{X}$ be arbitrary. Since R is reflexive, $(x, x) \in R$. So, $\exists R \in \boldsymbol{X}\big((x, x) \in R\big)$. Therefore, $(x, x) \in \bigcup \boldsymbol{X}$. Since $x \in A$ was arbitrary, $\bigcup \boldsymbol{X}$ is reflexive.

52. $\bigcap \boldsymbol{X}$ is symmetric on A.

True: Let $(x, y) \in \bigcap \boldsymbol{X}$ and let $R \in \boldsymbol{X}$. Then $(x, y) \in R$. Since R is an equivalence relation, R is symmetric. Therefore, $(y, x) \in R$. Since $R \in \boldsymbol{X}$ was arbitrary, $\forall R \in \boldsymbol{X}\big((y, x) \in R\big)$. So, $(y, x) \in \bigcap \boldsymbol{X}$. Since $(x, y) \in \bigcap \boldsymbol{X}$ was arbitrary, $\bigcap \boldsymbol{X}$ is symmetric.

53. $\bigcup \boldsymbol{X}$ is symmetric on A.

True: Let $(x, y) \in \bigcup \boldsymbol{X}$. Then there is $R \in \boldsymbol{X}$ with $(x, y) \in R$. Since R is an equivalence relation, R is symmetric. Therefore, $(y, x) \in R$. So, $\exists R \in \boldsymbol{X}\big((y, x) \in R\big)$. Therefore, $(y, x) \in \bigcup \boldsymbol{X}$. Since $(x, y) \in \bigcup \boldsymbol{X}$ was arbitrary, $\bigcup \boldsymbol{X}$ is symmetric.

54. $\bigcap \boldsymbol{X}$ is transitive on A.

True: Let $(x, y), (y, z) \in \bigcap \boldsymbol{X}$ and let $R \in \boldsymbol{X}$. Then $(x, y), (y, z) \in R$. Since R is an equivalence relation, R is transitive. Therefore, $(x, z) \in R$. Since $R \in \boldsymbol{X}$ was arbitrary, $\forall R \in \boldsymbol{X}\big((x, z) \in R\big)$. So, $(x, z) \in \bigcap \boldsymbol{X}$. Since $(x, y), (y, z) \in \bigcap \boldsymbol{X}$ was arbitrary, $\bigcap \boldsymbol{X}$ is transitive.

55. $\bigcup \boldsymbol{X}$ is transitive on A.

False: Let $A = \{0, 1, 2\}$.

Let $R = \{(0,0), (0,1), (1,0), (1,1), (2,2)\}$ and $S = \{(0,0), (1,1), (1,2), (2,1), (2,2)\}$. Then R and S are both equivalence relations.

$R \cup S = \{(0,0), (0,1), (1,0), (1,1), (1,2), (2,1), (2,2)\}$.

We have $(0,1), (1,2) \in R \cup S$, but $(0,2) \notin R \cup S$. So, $R \cup S$ is **not** transitive.

56. $\bigcap X$ is an equivalence relation on A.

True: Since $\bigcap X$ is reflexive, symmetric, and transitive, $\bigcap X$ is an equivalence relation.

57. $\bigcup X$ is an equivalence relation on A.

False: By Problem 55, $\bigcup X$ is not necessarily transitive on A.

LEVEL 5

Recall that we define the ordering $<_{\mathbb{Z}}$ on $\mathbb{Z}$ by $[(a,b)] <_{\mathbb{Z}} [(c,d)]$ if and only if $a + d <_{\mathbb{N}} b + c$, where $<_{\mathbb{N}}$ is the usual ordering on $\mathbb{N}$.

58. Show that $<_{\mathbb{Z}}$ is a well-defined relation on $\mathbb{Z}$.

Suppose that $(a,b) \sim (a',b')$ and $(c,d) \sim (c',d)$. Since $(a,b) \sim (a',b')$, $a + b' = b + a'$. Since $(c,d) \sim (c',d')$, $c + d' = d + c'$.

We need to check that $[(a,b)] <_{\mathbb{Z}} [(c,d)]$ if and only if $[(a',b')] <_{\mathbb{Z}} [(c',d')]$. We have

$[(a,b)] <_{\mathbb{Z}} [(c,d)]$ if and only if $a + d <_{\mathbb{N}} b + c$ if and only if $a + d + b' + c' <_{\mathbb{N}} b + c + b' + c'$ if and only if $a + b' + d + c' <_{\mathbb{N}} b + c + b' + c'$ if and only if $b + a' + c + d' <_{\mathbb{N}} b + c + b' + c'$ if and only if $b + c + a' + d' <_{\mathbb{N}} b + c + b' + c'$ if and only if $a' + d' < b' + c'$ if and only if $[(a',b')] <_{\mathbb{Z}} [(c',d')]$, as desired.

59. Show that $<_{\mathbb{Z}}$ is transitive on $\mathbb{Z}$.

Suppose that $[(a,b)] <_{\mathbb{Z}} [(c,d)]$ and $[(c,d)] <_{\mathbb{Z}} [(e,f)]$. Then $a + d <_{\mathbb{N}} b + c$ and $c + f <_{\mathbb{N}} d + e$. By adding each side of these two inequalities we get the inequality $a + d + c + f <_{\mathbb{N}} b + c + d + e$. Cancelling c and d from each side of this last inequality yields $a + f <_{\mathbb{N}} b + e$. Therefore, $[(a,b)] <_{\mathbb{Z}} [(e,f)]$.

60. Show that $<_{\mathbb{Z}}$ is trichotomous on $\mathbb{Z}$.

Suppose $[(a,b)] \nless_{\mathbb{Z}} [(c,d)]$ and $[(a,b)] \neq [(c,d)]$. Then $a + d \nless_{\mathbb{N}} b + c$ and $a + d \neq b + c$. Since trichotomy holds for $<_{\mathbb{N}}$, we have $b + c <_{\mathbb{N}} a + d$, or equivalently, $c + b <_{\mathbb{N}} d + a$. Therefore, $[(c,d)] <_{\mathbb{Z}} [(a,b)]$.

61. Explain why $(\mathbb{Z}, <_{\mathbb{Z}})$ is an ordered set.

$<_{\mathbb{Z}}$ is a well-defined binary relation on $\mathbb{Z}$ that is transitive and trichotomous on $\mathbb{Z}$ by Problems 58 – 60 above. So, $<_{\mathbb{Z}}$ is a strict linear ordering of $\mathbb{Z}$. Equivalently, $(\mathbb{Z}, <_{\mathbb{Z}})$ is an ordered set.

Let $R = \{(x,y) \in \mathbb{R} \times \mathbb{R} \mid x - y \in \mathbb{Z}\}$.

62. Show that R is reflexive on $\mathbb{R}$.

If $x \in \mathbb{R}$, then $x - x = 0 \in \mathbb{Z}$. So, xRx, and therefore, R is reflexive on $\mathbb{R}$.

63. Show that R is symmetric on $\mathbb{R}$.

If xRy, then $x - y \in \mathbb{Z}$. It follows that $y - x = -(x - y) \in \mathbb{Z}$. So, yRx, and therefore, R is symmetric on $\mathbb{R}$.

64. Show that R is transitive on $\mathbb{R}$.

If xRy and yRz, then $x - y \in \mathbb{Z}$ and $y - z \in \mathbb{Z}$. It follows that $x - z = (x - y) + (y - z) \in \mathbb{Z}$ (because the sum of two integers is an integer). So, xRy, and therefore, R is transitive.

65. Is R an equivalence relation on $\mathbb{R}$?

Yes. Since R is reflexive, symmetric, and transitive (by Problems 62 – 64), R is an equivalence relation.

Let $\sim$ be an equivalence relation on a set S.

66. Show that $\bigcup\{[x] \mid x \in S\} \subseteq S$.

Let $y \in \bigcup\{[x] \mid x \in S\}$. Then there is $x \in S$ with $y \in [x]$. By definition of $[x]$, $y \in S$. Therefore, $\bigcup\{[x] \mid x \in S\} \subseteq S$.

67. Show that $S \subseteq \bigcup\{[x] \mid x \in S\}$.

Let $y \in S$. Since $\sim$ is an equivalence relation, $y \sim y$. So, $y \in [y]$. Thus, $y \in \bigcup\{[x] \mid x \in S\}$. So, we have $S \subseteq \bigcup\{[x] \mid x \in S\}$.

68. Suppose that $x, y \in S$ and $[x] \cap [y] \neq \emptyset$. Show that $[x] \subseteq [y]$.

Since $[x] \cap [y] \neq \emptyset$, we can find $z \in [x] \cap [y]$. Then $x \sim z$ and $y \sim z$. Since $\sim$ is symmetric, $z \sim y$. Since $\sim$ is transitive, $x \sim y$. Since $\sim$ is symmetric, $y \sim x$.

Let $w \in [x]$. Then $x \sim w$. By transitivity, $y \sim w$. So, $w \in [y]$. Since $w \in [x]$ was arbitrary, $[x] \subseteq [y]$.

69. Suppose that $x, y \in S$ and $[x] \cap [y] \neq \emptyset$. Explain why $[x] = [y]$.

By Problem 68, $[x] \subseteq [y]$. By a symmetric argument to that given in the solution to Problem 68, $[y] \subseteq [x]$.

Since $[x] \subseteq [y]$ and $[y] \subseteq [x]$, we have $[x] = [y]$.

70. Show that the equivalence classes of $\sim$ form a partition of S.

By Problems 66 and 67, $\bigcup\{[x] \mid x \in S\} = S$. By Problems 68 and 69, each pair of equivalence classes is either disjoint or equal. Therefore, the set of equivalence classes partitions S.

Let $\boldsymbol{P}$ be a partition of S, and define the relation $\sim$ by $x \sim y$ if and only if there is $X \in \boldsymbol{P}$ with $x, y \in X$.

71. Show that $\sim$ is reflexive on S.

Let $x \in S$. Since $\boldsymbol{P}$ is a partition of S, $S = \bigcup \boldsymbol{P}$. So, there is $X \in \boldsymbol{P}$ with $x \in X$. It follows that $x \sim x$. Therefore, $\sim$ is reflexive.

72. Show that $\sim$ is symmetric on S.

If $x \sim y$, then there is $X \in \boldsymbol{P}$ with $x, y \in X$. So, $y, x \in X$ (obviously!). Thus, $y \sim x$, and therefore, $\sim$ is symmetric.

73. Show that $\sim$ is transitive on S.

If $x \sim y$ and $y \sim z$, then there are $X, Y \in \boldsymbol{P}$ with $x, y \in X$ and $y, z \in Y$. Since $y \in X$ and $y \in Y$, we have $y \in X \cap Y$. Since $\boldsymbol{P}$ is a partition and $X \cap Y \neq \emptyset$, we must have $X = Y$. So, $z \in X$. Thus, $x, z \in X$, and therefore, $x \sim z$. So, $\sim$ is transitive.

74. Show that $\sim$ is an equivalence relation on S.

By Problems 71 – 73, $\sim$ is reflexive, symmetric, and transitive on S. Therefore, $\sim$ is an equivalence relation on S.

75. Show that $\boldsymbol{P} \subseteq \{[x] \mid x \in S\}$.

Let $X \in \boldsymbol{P}$ and let $x \in X$. We show that $X = [x]$.

Let $y \in X$. Since $x, y \in X$, $x \sim y$. So $y \in [x]$. Thus, $X \subseteq [x]$.

Now, let $y \in [x]$. Then $x \sim y$. So, there is $Y \in \boldsymbol{P}$ such that $x, y \in Y$. Since $x \in X$ and $x \in Y$, $x \in X \cap Y$. Since $\boldsymbol{P}$ is a partition and $X \cap Y \neq \emptyset$, we must have $X = Y$. So, $y \in X$. Thus, $[x] \subseteq X$.

Since $X \subseteq [x]$ and $[x] \subseteq X$, we have $X = [x]$.

Since $X \in \boldsymbol{P}$ was arbitrary, we have shown $\boldsymbol{P} \subseteq \{[x] \mid x \in S\}$.

76. Show that $\{[x] \mid x \in S\} \subseteq \boldsymbol{P}$.

Let $X \in \{[x] \mid x \in S\}$. Then there is $x \in S$ such that $X = [x]$. Since $\boldsymbol{P}$ is a partition of S, $S = \bigcup \boldsymbol{P}$. So, there is $Y \in \boldsymbol{P}$ with $x \in Y$. We will show that $X = Y$.

Let $y \in X$. Then $x \sim y$. So, there is $Z \in \boldsymbol{P}$ with $x, y \in Z$. Since $x \in Y$ and $x \in Z$, $x \in Y \cap Z$. Since $\boldsymbol{P}$ is a partition and $Y \cap Z \neq \emptyset$, we must have $Y = Z$. So, $y \in Y$. Since $y \in X$ was arbitrary, $X \subseteq Y$.

Now, let $y \in Y$. Then $x \sim y$. So, $y \in [x] = X$. Since $y \in Y$ was arbitrary, $Y \subseteq X$.

Since $X \subseteq Y$ and $Y \subseteq X$, we have $X = Y$. Therefore, $X \in \boldsymbol{P}$.

Since $X \in \{[x] \mid x \in S\}$ was arbitrary, we have $\{[x] \mid x \in S\} \subseteq \boldsymbol{P}$.

77. Show that $\boldsymbol{P} = \{[x] \mid x \in S\}$.

By Problem 75, $\boldsymbol{P} \subseteq \{[x] \mid x \in S\}$. By Problem 76, $\{[x] \mid x \in S\} \subseteq \boldsymbol{P}$. Therefore, $\boldsymbol{P} = \{[x] \mid x \in S\}$, as desired.

Problem Set 5

LEVEL 1

Determine if each of the following relations is a function. State the domain and range of each such function.

1. $\{(0,0),(1,0)\}$

Function; domain $= \{0,1\}$, range $= \{0\}$

2. $\{(0,0),(0,1)\}$

Not a function

3. $\{(a,a),(b,b),(c,c)\}$

Function; domain $= \{a,b,c\}$, range $= \{a,b,c\}$

4. $\{(a,a),(a,b),(a,c)\}$

Not a function

5. $\{(a,a),(b,a),(c,a)\}$

Function; domain $= \{a,b,c\}$, range $= \{a\}$

6. $\{(a,b),(b,b),(c,d),(e,a)\}$

Function; domain $= \{a,b,c,e\}$, range $= \{a,b,d\}$

7. $\{(a,a),(a,b),(b,a)\}$

Not a function

8. $\{(0,0),(1,1),(2,2),(3,3),(4,4)\}$

Function; domain $= \{0,1,2,3,4\}$, range $= \{0,1,2,3,4\}$

9. $\{(0,a),(a,0),(b,1),(1,c),(c,0)\}$

Function; domain $= \{0,1,a,b,c\}$, range $= \{0,1,a,c\}$

10. $\{(0,a),(a,0),(0,1),(b,2)\}$

Not a function

Determine if each of the following functions is injective.

11. $\{(s, 0), (t, 0)\}$

Not injective

12. $\{(s, 0), (t, 1)\}$

Injective

13. $\{(a, a), (b, b), (c, c)\}$

Injective

14. $\{(a, a), (b, a), (c, b)\}$

Not injective

15. $\{(s, s), (t, s), (u, s)\}$

Not injective

16. $\{(a, b), (b, a), (c, d), (d, c)\}$

Injective

LEVEL 2

Determine if each of the following relations is a function. State the domain and range of each such function.

17. $\{(a, b) \in \mathbb{R}^2 \mid b \leq 0 \wedge a^2 + b^2 = 9\}$

Function; domain $= [-3, 3]$, range $= [-3, 0]$

18. $\{(a, b) \in \mathbb{R}^2 \mid a < 0 \wedge a^2 + b^2 = 9\}$

Not a function

19. $\left\{\left(\frac{a}{b}, c\right) \in \mathbb{Q} \times \mathbb{Z} \,\middle|\, c = a - b\right\}$

Not a function (for example, $\left(\frac{1}{2}, -1\right), \left(\frac{2}{4}, -2\right)$ are both in the set and $\frac{1}{2} = \frac{2}{4}$).

20. $\{(a + bi, c) \in \mathbb{C} \times \mathbb{R} \mid c = a\}$

Function; domain $= \mathbb{C}$, range $= \mathbb{R}$

21. $\{(c, a+bi) \in \mathbb{R} \times \mathbb{C} \mid c = \max\{a, b\}\}$, where $\max\{a, b\}$ is the larger of a or b.

Not a function (for example, $(2, 1+2i), (2, 2+i)$ are both in the set).

Determine if each of the following functions is injective, surjective, both, or neither.

22. $\{(a, b) \in \mathbb{R}^2 \mid b > 0 \wedge a^2 + b^2 = 1\}$

Neither

23. $\{(a, b) \in \mathbb{R} \times (0, 1] \mid b > 0 \wedge a^2 + b^2 = 1\}$

Surjective

24. $\{(a+bi, c) \in \mathbb{C} \times \mathbb{R} \mid c = a\}$

Surjective

25. $\{(n, m) \in \mathbb{N}^2 \mid m = 5n\}$

Injective

26. $\{(n, z) \in \mathbb{Z} \times \mathbb{C} \mid z = n + ni\}$

Injective

If $f: X \to Y$ and $A \subseteq X$, then the **image of A under f** is the set $f[A] = \{f(x) \mid x \in A\}$. Let $f: \{a, b, c, d\} \to \{0, 1, 2\}$ be defined by $f = \{(a, 0), (b, 0), (c, 1), (d, 2)\}$. Compute $f[A]$ for each of the following sets A:

27. $A = \{a\}$

$f[A] = f[\{a\}] = \{\mathbf{0}\}$

28. $A = \{a, b\}$

$f[A] = f[\{a, b\}] = \{\mathbf{0}\}$

29. $A = \{a, b, c\}$

$f[A] = f[\{a, b, c\}] = \{\mathbf{0}, \mathbf{1}\}$

30. $A = \{a, c\}$

$f[A] = f[\{a, c\}] = \{\mathbf{0}, \mathbf{1}\}$

31. $A = \{b, c, d\}$

$f[A] = f[\{b, c, d\}] = \{\mathbf{0}, \mathbf{1}, \mathbf{2}\}$

If $B \subseteq Y$, then the **inverse image of B under f** is the set $f^{-1}[B] = \{x \in X \mid f(x) \in B\}$. Let $f: \{a, b, c, d\} \to \{0, 1, 2\}$ be defined by $f = \{(a, 0), (b, 0), (c, 1), (d, 2)\}$. Compute $f^{-1}[B]$ for each of the following sets B:

32. $B = \{0\}$

$f^{-1}[B] = f^{-1}[\{0\}] = \{\boldsymbol{a}, \boldsymbol{b}\}$

33. $B = \{1\}$

$f^{-1}[B] = f^{-1}[\{1\}] = \{\boldsymbol{c}\}$

34. $B = \{0, 1\}$

$f^{-1}[B] = f^{-1}[\{0, 1\}] = \{\boldsymbol{a}, \boldsymbol{b}, \boldsymbol{c}\}$

35. $B = \{0, 2\}$

$f^{-1}[B] = f^{-1}[\{0, 2\}] = \{\boldsymbol{a}, \boldsymbol{b}, \boldsymbol{d}\}$

36. $B = \{0, 1, 2\}$

$f^{-1}[B] = f^{-1}[\{0, 1, 2\}] = \{\boldsymbol{a}, \boldsymbol{b}, \boldsymbol{c}, \boldsymbol{d}\}$

LEVEL 3

If $f: X \to Y$ and $A \subseteq X$, then the **image of A under f** is the set $f[A] = \{f(x) \mid x \in A\}$. Let $f: \mathbb{R} \to \mathbb{R}$ be defined by $f(x) = x^4$. Compute $f[A]$ for each of the following sets A:

37. $A = \mathbb{R}$

$f[A] = f[\mathbb{R}] = [\mathbf{0}, \infty)$

38. $A = \{-2, 0, 3\}$

$f[A] = f[\{-2, 0, 3\}] = \{\mathbf{0}, \mathbf{16}, \mathbf{81}\}$

39. $A = (-3, 2]$

$f[A] = f[(-3, 2]] = [\mathbf{0}, \mathbf{81})$

40. $A = (4, \infty)$

$f[A] = f[(4, \infty)] = (\mathbf{256}, \infty)$

41. $A = (-\infty, 1]$

$f[A] = f\big[(-\infty, 1]\big] = \mathbf{[0, \infty)}$

If $B \subseteq Y$, then the **inverse image of B under f** is the set $f^{-1}[B] = \{x \in X \mid f(x) \in B\}$. Let $f: \mathbb{R} \to \mathbb{R}$ be defined by $f(x) = x^4$. Compute $f^{-1}[B]$ for each of the following:

42. $B = \{16\}$

$f^{-1}[B] = f^{-1}[\{16\}] = \{\mathbf{-2, 2}\}$

43. $B = \{-1\}$

$f^{-1}[B] = f^{-1}[\{-1\}] = \emptyset$

44. $B = [0, \infty)$

$f^{-1}[B] = f^{-1}\big[[0, \infty)\big] = \mathbb{R}$

45. $B = (-\infty, 0)$

$f^{-1}[B] = f^{-1}[(-\infty, 0)] = \emptyset$

46. $B = (0, \infty)$

$f^{-1}[B] = f^{-1}[(0, \infty)] = (-\infty, 0) \cup (0, \infty)$

For each of the following sets X and Y, compute ${}^XY$:

47. $X = \emptyset, Y = \emptyset$

${}^XY = \emptyset$

48. $X = \{a\}, Y = \{b\}$

${}^XY = \big\{\{(\boldsymbol{a}, \boldsymbol{b})\}\big\}$

49. $X = \{a, b\}, Y = \{c\}$

${}^XY = \big\{\{(\boldsymbol{a}, \boldsymbol{c}), (\boldsymbol{b}, \boldsymbol{c})\}\big\}$

50. $X = \{a, b\}, Y = \{a\}$

${}^XY = \big\{\{(\boldsymbol{a}, \boldsymbol{a}), (\boldsymbol{b}, \boldsymbol{a})\}\big\}$

51. $X = \{a\}, Y = \{0, 1\}$

${}^XY = \big\{\{(\boldsymbol{a}, \mathbf{0})\}, \{(\boldsymbol{a}, \mathbf{1})\}\big\}$

52. $X = \{a, b\}, Y = \{a, b, c\}$

${}^{X}Y = \{\{(\boldsymbol{a}, \boldsymbol{a}), (\boldsymbol{b}, \boldsymbol{a})\}, \{(\boldsymbol{a}, \boldsymbol{a}), (\boldsymbol{b}, \boldsymbol{b})\}, \{(\boldsymbol{a}, \boldsymbol{a}), (\boldsymbol{b}, \boldsymbol{c})\}, \{(\boldsymbol{a}, \boldsymbol{b}), (\boldsymbol{b}, \boldsymbol{a})\}, \{(\boldsymbol{a}, \boldsymbol{b}), (\boldsymbol{b}, \boldsymbol{b})\}, \{(\boldsymbol{a}, \boldsymbol{b}), (\boldsymbol{b}, \boldsymbol{c})\},$
$\{(\boldsymbol{a}, \boldsymbol{c}), (\boldsymbol{b}, \boldsymbol{a})\}, \{(\boldsymbol{a}, \boldsymbol{c}), (\boldsymbol{b}, \boldsymbol{b})\}, \{(\boldsymbol{a}, \boldsymbol{c}), (\boldsymbol{b}, \boldsymbol{c})\}\}$

53. $X = \{a, b, c\}, Y = \{0, 1\}$

${}^{X}Y = \{\{(\boldsymbol{a}, \boldsymbol{0}), (\boldsymbol{b}, \boldsymbol{0}), (\boldsymbol{c}, \boldsymbol{0})\}, \{(\boldsymbol{a}, \boldsymbol{0}), (\boldsymbol{b}, \boldsymbol{0}), (\boldsymbol{c}, \boldsymbol{1})\}, \{(\boldsymbol{a}, \boldsymbol{0}), (\boldsymbol{b}, \boldsymbol{1}), (\boldsymbol{c}, \boldsymbol{0})\}, \{(\boldsymbol{a}, \boldsymbol{0}), (\boldsymbol{b}, \boldsymbol{1}), (\boldsymbol{c}, \boldsymbol{1})\}$
$\{(\boldsymbol{a}, \boldsymbol{1}), (\boldsymbol{b}, \boldsymbol{0}), (\boldsymbol{c}, \boldsymbol{0})\}, \{(\boldsymbol{a}, \boldsymbol{1}), (\boldsymbol{b}, \boldsymbol{0}), (\boldsymbol{c}, \boldsymbol{1})\}, \{(\boldsymbol{a}, \boldsymbol{1}), (\boldsymbol{b}, \boldsymbol{1}), (\boldsymbol{c}, \boldsymbol{0})\}, \{(\boldsymbol{a}, \boldsymbol{1}), (\boldsymbol{b}, \boldsymbol{1}), (\boldsymbol{c}, \boldsymbol{1})\}\}$

54. $X = \{a, b, c\}, Y = \{0, 1, 2\}$

${}^{X}Y = \{\{(\boldsymbol{a}, \boldsymbol{0}), (\boldsymbol{b}, \boldsymbol{0}), (\boldsymbol{c}, \boldsymbol{0})\}, \{(\boldsymbol{a}, \boldsymbol{0}), (\boldsymbol{b}, \boldsymbol{0}), (\boldsymbol{c}, \boldsymbol{1})\}, \{(\boldsymbol{a}, \boldsymbol{0}), (\boldsymbol{b}, \boldsymbol{0}), (\boldsymbol{c}, \boldsymbol{2})\}, \{(\boldsymbol{a}, \boldsymbol{0}), (\boldsymbol{b}, \boldsymbol{1}), (\boldsymbol{c}, \boldsymbol{0})\},$
$\{(\boldsymbol{a}, \boldsymbol{0}), (\boldsymbol{b}, \boldsymbol{1}), (\boldsymbol{c}, \boldsymbol{1})\}, \{(\boldsymbol{a}, \boldsymbol{0}), (\boldsymbol{b}, \boldsymbol{1}), (\boldsymbol{c}, \boldsymbol{2})\}, \{(\boldsymbol{a}, \boldsymbol{0}), (\boldsymbol{b}, \boldsymbol{2}), (\boldsymbol{c}, \boldsymbol{0})\}, \{(\boldsymbol{a}, \boldsymbol{0}), (\boldsymbol{b}, \boldsymbol{2}), (\boldsymbol{c}, \boldsymbol{1})\},$
$\{(\boldsymbol{a}, \boldsymbol{0}), (\boldsymbol{b}, \boldsymbol{2}), (\boldsymbol{c}, \boldsymbol{2})\}, \{(\boldsymbol{a}, \boldsymbol{1}), (\boldsymbol{b}, \boldsymbol{0}), (\boldsymbol{c}, \boldsymbol{0})\}, \{(\boldsymbol{a}, \boldsymbol{1}), (\boldsymbol{b}, \boldsymbol{0}), (\boldsymbol{c}, \boldsymbol{1})\}, \{(\boldsymbol{a}, \boldsymbol{1}), (\boldsymbol{b}, \boldsymbol{0}), (\boldsymbol{c}, \boldsymbol{2})\},$
$\{(\boldsymbol{a}, \boldsymbol{1}), (\boldsymbol{b}, \boldsymbol{1}), (\boldsymbol{c}, \boldsymbol{0})\}, \{(\boldsymbol{a}, \boldsymbol{1}), (\boldsymbol{b}, \boldsymbol{1}), (\boldsymbol{c}, \boldsymbol{1})\}, \{(\boldsymbol{a}, \boldsymbol{1}), (\boldsymbol{b}, \boldsymbol{1}), (\boldsymbol{c}, \boldsymbol{2})\}, \{(\boldsymbol{a}, \boldsymbol{1}), (\boldsymbol{b}, \boldsymbol{2}), (\boldsymbol{c}, \boldsymbol{0})\},$
$\{(\boldsymbol{a}, \boldsymbol{1}), (\boldsymbol{b}, \boldsymbol{2}), (\boldsymbol{c}, \boldsymbol{1})\}, \{(\boldsymbol{a}, \boldsymbol{1}), (\boldsymbol{b}, \boldsymbol{2}), (\boldsymbol{c}, \boldsymbol{2})\}, \{(\boldsymbol{a}, \boldsymbol{2}), (\boldsymbol{b}, \boldsymbol{0}), (\boldsymbol{c}, \boldsymbol{0})\}, \{(\boldsymbol{a}, \boldsymbol{2}), (\boldsymbol{b}, \boldsymbol{0}), (\boldsymbol{c}, \boldsymbol{1})\},$
$\{(\boldsymbol{a}, \boldsymbol{2}), (\boldsymbol{b}, \boldsymbol{0}), (\boldsymbol{c}, \boldsymbol{2})\}, \{(\boldsymbol{a}, \boldsymbol{2}), (\boldsymbol{b}, \boldsymbol{1}), (\boldsymbol{c}, \boldsymbol{0})\}, \{(\boldsymbol{a}, \boldsymbol{2}), (\boldsymbol{b}, \boldsymbol{1}), (\boldsymbol{c}, \boldsymbol{1})\}, \{(\boldsymbol{a}, \boldsymbol{2}), (\boldsymbol{b}, \boldsymbol{1}), (\boldsymbol{c}, \boldsymbol{2})\},$
$\{(\boldsymbol{a}, \boldsymbol{2}), (\boldsymbol{b}, \boldsymbol{2}), (\boldsymbol{c}, \boldsymbol{0})\}, \{(\boldsymbol{a}, \boldsymbol{2}), (\boldsymbol{b}, \boldsymbol{2}), (\boldsymbol{c}, \boldsymbol{1})\}, \{(\boldsymbol{a}, \boldsymbol{2}), (\boldsymbol{b}, \boldsymbol{2}), (\boldsymbol{c}, \boldsymbol{2})\}\}$

LEVEL 4

For $f, g \in {}^{\mathbb{R}}\mathbb{R}$, define $f \preccurlyeq g$ if and only if for all $x \in \mathbb{R}$, $f(x) \leq g(x)$.

55. Is $\preccurlyeq$ reflexive on ${}^{\mathbb{R}}\mathbb{R}$?

Yes

56. Is $\preccurlyeq$ symmetric on ${}^{\mathbb{R}}\mathbb{R}$?

No

57. Is $\preccurlyeq$ transitive on ${}^{\mathbb{R}}\mathbb{R}$?

Yes

58. Is $\preccurlyeq$ antireflexive on ${}^{\mathbb{R}}\mathbb{R}$?

No

59. Is $\preccurlyeq$ antisymmetric on ${}^{\mathbb{R}}\mathbb{R}$?

Yes

60. Is $\preccurlyeq$ trichotomous on ${}^{\mathbb{R}}\mathbb{R}$?

No

61. Does $\preccurlyeq$ satisfy the comparability condition on ${}^{\mathbb{R}}\mathbb{R}$?

No (for example, $f(x) = x$ and $g(x) = x^2$ are incomparable).

62. Is $\preccurlyeq$ a partial ordering on ${}^{\mathbb{R}}\mathbb{R}$?

Yes (because $\preccurlyeq$ is reflexive, antisymmetric, and transitive on ${}^{\mathbb{R}}\mathbb{R}$).

63. Is $\preccurlyeq$ a linear ordering on ${}^{\mathbb{R}}\mathbb{R}$?

No (because $\preccurlyeq$ does not satisfy the comparability condition on ${}^{\mathbb{R}}\mathbb{R}$).

For $f, g \in {}^{\mathbb{R}}\mathbb{R}$, define $f \preccurlyeq^* g$ if and only if there is an $x \in \mathbb{R}$ such that $f(x) \leq g(x)$.

64. Is $\preccurlyeq^*$ reflexive on ${}^{\mathbb{R}}\mathbb{R}$?

Yes

65. Is $\preccurlyeq^*$ symmetric on ${}^{\mathbb{R}}\mathbb{R}$?

No

66. Is $\preccurlyeq^*$ transitive on ${}^{\mathbb{R}}\mathbb{R}$?

No (for example, let $f(x) = \begin{cases} 0 & \text{if } x \neq 1 \\ 5 & \text{if } x = 1 \end{cases}$, $g(x) = 1$, and $h(x) = \begin{cases} -1 & \text{if } x \neq 1. \\ 2 & \text{if } x = 1. \end{cases}$ Then $f \preccurlyeq^* g$ and $g \preccurlyeq^* h$, but $f \not\preccurlyeq^* h$.

67. Is $\preccurlyeq^*$ antireflexive on ${}^{\mathbb{R}}\mathbb{R}$?

No

68. Is $\preccurlyeq^*$ antisymmetric on ${}^{\mathbb{R}}\mathbb{R}$?

No (for example, $f(x) = x$ and $g(x) = x^2$ satisfy $f \preccurlyeq^* g$ and $g \preccurlyeq^* f$, but $f \neq g$).

69. Is $\preccurlyeq^*$ trichotomous on ${}^{\mathbb{R}}\mathbb{R}$?

No

70. Does $\preccurlyeq^*$satisfy the comparability condition on ${}^{\mathbb{R}}\mathbb{R}$?

Yes

71. Is $\preccurlyeq^*$ a partial ordering on ${}^{\mathbb{R}}\mathbb{R}$?

No (because $\preccurlyeq^*$ is not antisymmetric or transitive on ${}^{\mathbb{R}}\mathbb{R}$).

72. Is $\preccurlyeq^*$ a linear ordering on ${}^{\mathbb{R}}\mathbb{R}$?

No (because $\preccurlyeq^*$ is not a partial ordering on ${}^{\mathbb{R}}\mathbb{R}$).

LEVEL 5

Determine if each of the following sequences is a Cauchy sequence. Are any of the Cauchy sequences equivalent?

73. $(x_n) = \left(1 + \frac{1}{n+1}\right)$

Cauchy sequence

74. $(y_n) = (2^n)$

Not a Cauchy sequence

75. $(z_n) = \left(1 - \frac{1}{2n+1}\right)$

Cauchy sequence

The Cauchy sequences (x_n) and (z_n) are equivalent.

Let $f: A \to B$ and $g: B \to C$. Show that each of the following is true.

76. If f is injective and g is injective, then $g \circ f$ is injective.

Suppose that $f: A \hookrightarrow B$ and $g: B \hookrightarrow C$, and let $x, y \in A$ with $x \neq y$. Since f is injective, $f(x) \neq f(y)$. Since g is injective, $g(f(x)) \neq g(f(y)$. So, $(g \circ f)(x) \neq (g \circ f)(y)$. Since $x, y \in A$ were arbitrary, $g \circ f: A \hookrightarrow C$.

77. If f is surjective and g is surjective, then $g \circ f$ is surjective.

Suppose that $f: A \mapsto B$ and $g: B \mapsto C$, and let $c \in C$. Since g surjective, there is $b \in B$ with $g(b) = c$. Since f is surjective, there is $a \in A$ with $f(a) = b$. So, $(g \circ f)(a) = g(f(a)) = g(b) = c$. Since $c \in C$ was arbitrary, $g \circ f$ is surjective

78. If f is bijective and g is bijective, then $g \circ f$ is bijective.

Suppose that $f: A \cong B$ and $g: B \cong C$. Then f and g are injective. By Problem 76, $g \circ f$ is injective. Also, f and g are surjective. By Problem 77, $g \circ f$ is surjective. Since $g \circ f$ is both injective and surjective, $g \circ f$ is bijective.

Let $f: A \cong B$. Show that each of the following is true.

79. If f is bijective, then $f^{-1} \circ f = i_A$.

Let $a \in A$ with $f(a) = b$. Then $f^{-1}(b) = a$, and so, $(f^{-1} \circ f)(a) = f^{-1}(f(a)) = f^{-1}(b) = a$. Since $i_A(a) = a$, we see that $(f^{-1} \circ f)(a) = i_A(a)$. Since $a \in A$ was arbitrary, $f^{-1} \circ f = i_A$.

80. If f is bijective, then $f \circ f^{-1} = i_B$.

Let $b \in B$. Since $f: A \cong B$, there is a unique $a \in A$ with $f(a) = b$. Equivalently, $f^{-1}(b) = a$. We have $(f \circ f^{-1})(b) = f(f^{-1}(b)) = f(a) = b$. Since $i_B(b) = b$, we see that $(f \circ f^{-1})(b) = i_B(b)$. Since $b \in B$ was arbitrary, $f \circ f^{-1} = i_B$.

Problem Set 6

LEVEL 1

For each of the following, determine if A and B are equinumerous. If so, provide a bijection from one set to the other. If not, explain why.

1. $A = \emptyset, B = \{\emptyset\}$

$\boldsymbol{A} \nsim \boldsymbol{B}$ because $|A| = 0$ and $|B| = 1$.

2. $A = \mathcal{P}(\emptyset), B = \{\emptyset\}$

$\boldsymbol{A} \sim \boldsymbol{B}$ because $A = B$.

3. $A = \{0, 1, 2\}, B = \{\Delta, \square, \perp\}$

$\boldsymbol{A} \sim \boldsymbol{B}$ via the bijection $\{(0, \Delta), (1, \square), (2, \perp)\}$.

4. $A = \{n \in \mathbb{N} \mid 5 \leq n \leq 250\}, B = \{6, 7, 8, \ldots, 250, 251\}$

$\boldsymbol{A} \sim \boldsymbol{B}$ via the bijection $f: A \to B$ defined by $f(n) = n + 1$.

5. $A = \{2n \mid n \in \mathbb{Z}\}, B = \{2n \in \mathbb{Z} \mid -100 < n < 9000\}$

$\boldsymbol{A} \nsim \boldsymbol{B}$ because A is infinite and B is finite.

6. $A = {}^{\mathbb{Z}}2, B = \mathcal{P}(\mathbb{Z})$

$\boldsymbol{A} \sim \boldsymbol{B}$ via the bijection $F: {}^{\mathbb{Z}}2 \to \mathcal{P}(\mathbb{Z})$ defined by $F(f) = \{n \in \mathbb{Z} \mid f(n) = 1\}$ (see Exercise 5.21).

7. $A = {}^{\mathbb{Q}}2, B = \mathcal{P}(\mathbb{Z})$ (you may use Problem 63 below)

$\boldsymbol{A} \sim \boldsymbol{B}$ because ${}^{\mathbb{Q}}2 \sim {}^{\mathbb{Z}}2 \sim \mathcal{P}(\mathbb{Z})$ (use part 4 of Example 6.1, Exercise 6.3, part 3 of Exercise 6.5, and Problem 63 below).

8. $A = (0, 1), B = (0, 2)$

$\boldsymbol{A} \sim \boldsymbol{B}$ via the bijection $f: (0, 1) \to (0, 2)$ defined by $f(x) = 2x$.

State whether each of the following sets is finite, denumerable, or uncountable.

9. $\{3, 4, 5, \ldots\}$

Denumerable

10. $\{3, 4, 5, \ldots, 888, 889\}$

Finite

11. $\mathcal{P}(\{3, 4, 5, \ldots\})$

Uncountable (the power set of an infinite set is uncountable)

12. $\mathcal{P}(\{3, 4, 5, \ldots, 888, 889\})$

Finite (the power set of a finite set is finite)

13. $\mathcal{P}(\mathcal{P}(\{3, 4, 5, \ldots, 888, 889\}))$

Finite (the power set of a finite set is finite)

14. $\bigcup\{\{n\} \mid n \in \mathbb{Z}\}$

Denumerable (this set is equal to $\mathbb{Z}$)

15. $\mathbb{Z} \setminus \mathbb{N}$

Denumerable

16. $\mathcal{P}(\mathbb{Q} \setminus \mathbb{Z})$

Uncountable ($\mathbb{Q} \setminus \mathbb{Z}$ is infinite and the power set of an infinite set is uncountable)

LEVEL 2

Show that each of the following sets is denumerable by generating a list.

17. $3\mathbb{N}$

$\mathbf{0, 3, 6, 9, 12, 15, \ldots}$

18. $3\mathbb{Z}$

$\mathbf{0, -3, 3, -6, 6, -9, 9, -12, 12, -15, 15, \ldots}$

19. $\{n \in \mathbb{N} \mid n \text{ is not divisible by } 5\}$

$\mathbf{1, 2, 3, 4, 6, 7, 8, 9, 11, 12, 13, 14, 16, \ldots}$

20. $\{n \in \mathbb{Z} \mid n \text{ is not divisible by } 5\}$

$\mathbf{-1, 1, -2, 2, -3, 3, -4, 4, -6, 6, -7, 7, -8, 8, -9, 9, -11, 11, -12, 12, \ldots}$

21. $\mathbb{Q}^-$

$-1, -2, -\frac{1}{2}, -3, -\frac{1}{3}, -4, -\frac{3}{2}, -\frac{2}{3}, -\frac{1}{4}, -5, -\frac{1}{5}, \ldots$

22. $\mathbb{N} \times \mathbb{N}$

$(\mathbf{0},\mathbf{0}),(\mathbf{0},\mathbf{1}),(\mathbf{1},\mathbf{0}),(\mathbf{0},\mathbf{2}),(\mathbf{1},\mathbf{1}),(\mathbf{1},\mathbf{0}),(\mathbf{0},\mathbf{3}),(\mathbf{1},\mathbf{2}),(\mathbf{2},\mathbf{1}),(\mathbf{3},\mathbf{0}),(\mathbf{0},\mathbf{4}),(\mathbf{1},\mathbf{3}),\ldots$

23. $\mathbb{Z} \times \mathbb{Z}$

$(\mathbf{0},\mathbf{0}),(\mathbf{0},-\mathbf{1}),(\mathbf{0},\mathbf{1}),(-\mathbf{1},\mathbf{0}),(\mathbf{1},\mathbf{0}),(\mathbf{0},-\mathbf{2}),(\mathbf{0},\mathbf{2}),(-\mathbf{1},-\mathbf{1}),(-\mathbf{1},\mathbf{1}),(\mathbf{1},-\mathbf{1}),(\mathbf{1},\mathbf{1}),(\mathbf{0},-\mathbf{3}),\ldots$

Use one of the symbols $\prec$ or $\sim$ to describe the relationship between A and B.

24. $A = \{0,1,2\}$, $B = \{1,2,3,4\}$

$\boldsymbol{A} \prec \boldsymbol{B}$ because $|A| = 3$ and $|B| = 4$.

25. $A = \{0,1,2\}$, $B = \mathcal{P}(\{0,1,2\})$

$\boldsymbol{A} \prec \boldsymbol{B}$ by Equinumerosity Fact 1 (Cantor's Theorem).

26. $A = \mathbb{Q}$, $B = \mathcal{P}(\mathbb{Q})$

$\boldsymbol{A} \prec \boldsymbol{B}$ by Equinumerosity Fact 1 (Cantor's Theorem).

27. $A = \mathcal{P}(\mathbb{N})$, $B = \mathcal{P}(\mathbb{Z})$ (use Problem 57 below)

$\boldsymbol{A} \sim \boldsymbol{B}$ by Problem 57 below and because $\mathbb{N} \sim \mathbb{Z}$.

28. $A = \mathcal{P}(\mathcal{P}(\mathbb{N}))$, $B = \mathcal{P}(\mathbb{Q})$ (use Problem 57 below)

$\boldsymbol{B} \prec \boldsymbol{A}$ because $\mathcal{P}(\mathbb{Q}) \sim \mathcal{P}(\mathbb{N})$ (by Problem 57 below and because $\mathbb{N} \sim \mathbb{Q}$) and because $\mathcal{P}(\mathbb{N}) \prec \mathcal{P}(\mathcal{P}(\mathbb{N}))$ (by Cantor's Theorem).

LEVEL 3

Let $f: \mathbb{N} \to \mathbb{Z}$ be defined by $f(n) = \begin{cases} \frac{n}{2} & \text{if } n \text{ is even.} \\ -\frac{n+1}{2} & \text{if } n \text{ is odd.} \end{cases}$

29. Show that f is injective.

Let $n, m \in \mathbb{N}$ with $f(n) = f(m)$. If n and m are both even, we have $\frac{n}{2} = \frac{m}{2}$, and so, $2 \cdot \frac{n}{2} = 2 \cdot \frac{m}{2}$. Thus, $n = m$. If n and m are both odd, we have $-\frac{n+1}{2} = -\frac{m+1}{2}$, and so, $\frac{n+1}{2} = \frac{m+1}{2}$. Thus, $2 \cdot \frac{n+1}{2} = 2 \cdot \frac{m+1}{2}$. So, $n + 1 = m + 1$, and therefore, $n = m$. If n is even and m is odd, then we have $\frac{n}{2} = -\frac{m+1}{2}$. So, $2 \cdot \frac{n}{2} = 2\left(-\frac{m+1}{2}\right)$. Therefore, $n = -(m+1)$. Since $m \in \mathbb{N}$, $m \geq 0$. So, $m + 1 \geq 1$. Therefore, $n = -(m+1) \leq -1$, which is impossible because natural numbers cannot be negative. So, it is not possible for n to be even, m to be odd, and $f(n) = f(m)$. Similarly, we cannot have n odd and m even. So, f is injective.

30. Show that f is surjective.

Let $k \in \mathbb{Z}$. If $k \geq 0$, then $2k \in \mathbb{N}$ and $f(2k) = \frac{2k}{2} = k$. If $k < 0$, then $-2k > 0$, and so, $-2k - 1 \in \mathbb{N}$. Then $f(-2k-1) = -\frac{(-2k-1)+1}{2} = -\frac{-2k}{2} = k$. So, f is surjective.

31. Show that f is bijective.

By Problem 29, f is injective. By Problem 30, f is surjective. So, f is bijective.

Note: It should really be checked that whenever $n \in \mathbb{N}$, $f(n) \in \mathbb{Z}$. To see this, first observe that if n is even, then there is $k \in \mathbb{Z}$ with $n = 2k$, and so, $\frac{n}{2} = \frac{2k}{2} = k \in \mathbb{Z}$. Now, if n is odd, there is $k \in \mathbb{Z}$ with $n = 2k+1$, and so, $-\frac{n+1}{2} = -\frac{(2k+1)+1}{2} = -\frac{2k+2}{2} = -\frac{2(k+1)}{2} = -(k+1) \in \mathbb{Z}$. So, f does in fact take each natural number to an integer.

Let $a, b \in \mathbb{R}$ with $a \neq 0$ and $a < b$. Define $f: [0,1] \to [a,b]$ by $f(x) = (b-a)x + a$.

32. Show that $f(0) = a$.

$f(0) = (b-a) \cdot 0 + a = 0 + a = a$.

33. Show that $f(1) = b$.

$f(1) = (b-a) \cdot 1 + a = b - a + a = b$.

34. Show that for any x with $0 < x < 1$, we have $a < f(x) < b$.

If $0 < x < 1$, then $(b-a) \cdot 0 < (b-a)x < (b-a) \cdot 1$, or equivalently, $0 < (b-a)x < b-a$. Therefore, $0 + a < (b-a)x + a < (b-a) + a$, or equivalently, $a < f(x) < b$.

Note: We first multiplied each part of the inequality $0 < x < 1$ by the positive number $b-a$. We then added a to each part of the inequality $0 < (b-a)x < b-a$.

35. Show that f is injective.

Let $x, y \in [0,1]$ and suppose that $f(x) = f(y)$. So, $(b-a)x + a = (b-a)y + a$. Subtracting a from each side of this equation yields $(b-a)x = (b-a)y$. Dividing each side of this last equation by the positive number $b-a$ yields $x = y$. Since $x, y \in [0,1]$ were arbitrary, f is injective.

36. Show that f is surjective.

Let $y \in [a,b]$ and let $x = \frac{y-a}{b-a}$. Since $a \leq y \leq b$, we have $0 \leq y - a \leq b - a$. Therefore, we have $\frac{0}{b-a} \leq \frac{y-a}{b-a} \leq \frac{b-a}{b-a}$, or equivalently, $0 \leq x \leq 1$. So, $x \in [0,1]$. Also, we have

$$f(x) = (b-a)\left(\frac{y-a}{b-a}\right) + a = (y-a) + a = y.$$

Since $y \in [a,b]$ was arbitrary, f is surjective.

37. Show that f is bijective.

By Problem 35, f is injective. By Problem 36, f is surjective. So, f is bijective.

The functions $f: \mathbb{R} \to (0, \infty)$ and $g: (0, \infty) \to (0, 1)$ defined by $f(x) = 2^x$ and $g(x) = \frac{1}{x^2+1}$ are bijections. Use this information to verify each of the following.

38. $\mathbb{R} \sim (0, \infty)$.

This is because f is a bijection.

39. $(0, \infty) \sim (0, 1)$.

This is because g is a bijection.

40. $(0, 1) \sim \mathbb{R}$.

By Problem 39, $(0, \infty) \sim (0, 1)$. Since $\sim$ is symmetric, $(0, 1) \sim (0, \infty)$. By Problem 38, $\mathbb{R} \sim (0, \infty)$. Since $\sim$ is symmetric, $(0, \infty) \sim \mathbb{R}$. Since $\sim$ is transitive, $(0, 1) \sim \mathbb{R}$.

41. $[0, 1] \sim \mathbb{R}$.

By Example 6.10, $(0, 1) \sim [0, 1]$. Since $\sim$ is symmetric, $[0, 1] \sim (0, 1)$. By Problem 40, $(0, 1) \sim \mathbb{R}$. Since $\sim$ is transitive, $[0, 1] \sim \mathbb{R}$.

42. $(0, 1] \sim \mathbb{R}$.

$(0, 1] \subseteq [0, 1] \sim \mathbb{R}$ (by Problem 41). So, $(0, 1] \preccurlyeq \mathbb{R}$. Now, by Problem 40, $(0, 1) \sim \mathbb{R}$. Since $\sim$ is symmetric, $\mathbb{R} \sim (0, 1)$. Therefore, we have $\mathbb{R} \sim (0, 1) \subseteq (0, 1]$. So, $\mathbb{R} \preccurlyeq (0, 1]$. We now invoke the Cantor-Shroeder-Bernstein Theorem to get $(0, 1] \sim \mathbb{R}$.

43. $[0, 1) \sim \mathbb{R}$.

$[0, 1) \subseteq [0, 1] \sim \mathbb{R}$ (by Problem 41). So, $[0, 1) \preccurlyeq \mathbb{R}$. Now, by Problem 40, $(0, 1) \sim \mathbb{R}$. Since $\sim$ is symmetric, $\mathbb{R} \sim (0, 1)$. Therefore, we have $\mathbb{R} \sim (0, 1) \subseteq [0, 1)$. So, $\mathbb{R} \preccurlyeq [0, 1)$. We now invoke the Cantor-Shroeder-Bernstein Theorem to get $(0, 1] \sim \mathbb{R}$.

44. $(-\infty, 1) \sim (0, \infty)$.

$(-\infty, 1) \subseteq \mathbb{R} \sim (0, \infty)$ (by Problem 38). So, $(-\infty, 1) \preccurlyeq (0, \infty)$. Now, by Problem 39, $(0, \infty) \sim (0, 1)$. Therefore, we have $(0, \infty) \sim (0, 1) \subseteq (-\infty, 1)$. So, $(0, \infty) \preccurlyeq (-\infty, 1)$. We now invoke the Cantor-Shroeder-Bernstein Theorem to get $(-\infty, 1) \sim (0, \infty)$.

LEVEL 4

Show that each of the following is true.

45. If $b > 0$, then $[0, 1] \sim [0, b]$.

The function $f: [0, 1] \to [0, b]$ given by $f(x) = bx$ is a bijection.

46. If $a \neq 0$ and $a < b$, then $[0, 1] \sim [a, b]$.

The function $f: [0, 1] \to [a, b]$ given by $f(x) = (b - a)x + a$ is a bijection (by Problem 37).

47. Any two bounded closed intervals are equinumerous.

Let $[a, b]$ and $[c, d]$ be bounded closed intervals. By Problems 45 and 46, $[0, 1] \sim [a, b]$ and $[0, 1] \sim [c, d]$. Since $\sim$ is an equivalence relation, $[a, b] \sim [c, d]$.

48. Any two bounded intervals are equinumerous.

By Problem 47, any two bounded closed intervals are equinumerous. So, it suffices to show that each bounded interval is equinumerous with a bounded closed interval.

We have $(a, b) \subseteq (a, b] \subseteq [a, b]$ and $[a, b) \subseteq [a, b]$. So, we have $(a, b) \preccurlyeq [a, b]$, $(a, b] \preccurlyeq [a, b]$, and $[a, b) \preccurlyeq [a, b]$.

For the other direction, observe that $\left[a + \frac{b-a}{4}, b - \frac{b-a}{4}\right] = \left[\frac{3a+b}{4}, \frac{a+3b}{4}\right] \subseteq (a, b)$. It follows from Problem 47 that $[a, b] \sim \left[\frac{3a+b}{4}, \frac{a+3b}{4}\right] \subseteq (a, b) \subseteq (a, b]$ and $[a, b] \sim \left[\frac{3a+b}{4}, \frac{a+3b}{4}\right] \subseteq (a, b) \subseteq [a, b)$. So, we have $[a, b] \preccurlyeq (a, b)$, $[a, b] \preccurlyeq (a, b]$, and $[a, b] \preccurlyeq [a, b)$

The result now follows from the Cantor-Shroeder-Bernstein Theorem.

Note: For the above argument to work, it is important that we have $a < \frac{3a+b}{4} < \frac{a+3b}{4} < b$. We see that this sequence of inequalities is equivalent to $4a < 3a + b < a + 3b < 4b$ (just multiply each part by 4). Now, $4a < 3a + b$ is equivalent to $a < b$ (by subtracting $3a$), which is true. Similarly, we see that $3a + b < a + 3b$ is equivalent to $2a < 2b$ (by subtracting a and then subtracting b), which in turn is equivalent to $a < b$ (by dividing by 2). Finally, $a + 3b < 4b$ is also equivalent to $a < b$ (by subtracting $3b$).

49. Any two intervals are equinumerous (including $\mathbb{R}$ itself).

By Problems 41 and 48 all bounded intervals are equinumerous with $\mathbb{R}$.

We also have the following.

$$(a, \infty) \subseteq [a, \infty) \subseteq \mathbb{R} \sim (a, a + 1) \subseteq (a, \infty)$$
$$(-\infty, b) \subseteq (-\infty, b] \subseteq \mathbb{R} \sim (b - 1, b) \subseteq (-\infty, b)$$

Therefore, all unbounded intervals are equinumerous with $\mathbb{R}$

It follows that any two intervals of real numbers are equinumerous.

Show that each of the following is true.

50. ${}^{\mathbb{N}}\mathbb{N} \preccurlyeq \mathcal{P}(\mathbb{N})$. (you may use Problems 53 and 57 below)

${}^{\mathbb{N}}\mathbb{N} \subseteq \mathcal{P}(\mathbb{N} \times \mathbb{N})$ by the definition of ${}^{\mathbb{N}}\mathbb{N}$. So, ${}^{\mathbb{N}}\mathbb{N} \preccurlyeq \mathcal{P}(\mathbb{N} \times \mathbb{N})$. By Problem 22, $\mathbb{N} \times \mathbb{N} \sim \mathbb{N}$. So, by Problem 57 below, $\mathcal{P}(\mathbb{N} \times \mathbb{N}) \sim \mathcal{P}(\mathbb{N})$. So, $\mathcal{P}(\mathbb{N} \times \mathbb{N}) \preccurlyeq \mathcal{P}(\mathbb{N})$. Since $\preccurlyeq$ is transitive (by Problem 53 below), ${}^{\mathbb{N}}\mathbb{N} \preccurlyeq \mathcal{P}(\mathbb{N})$.

51. $\mathcal{P}(\mathbb{N}) \preccurlyeq {}^{\mathbb{N}}\mathbb{N}$. (you may use Problem 53 below)

$\mathcal{P}(\mathbb{N}) \sim {}^{\mathbb{N}}\{0,1\}$. So, $\mathcal{P}(\mathbb{N}) \preccurlyeq {}^{\mathbb{N}}\{0,1\}$. Also, ${}^{\mathbb{N}}\{0,1\} \subseteq {}^{\mathbb{N}}\mathbb{N}$, and so, ${}^{\mathbb{N}}\{0,1\} \preccurlyeq {}^{\mathbb{N}}\mathbb{N}$. Since $\preccurlyeq$ is transitive, $\mathcal{P}(\mathbb{N}) \preccurlyeq {}^{\mathbb{N}}\mathbb{N}$.

52. ${}^{\mathbb{N}}\mathbb{N} \sim \mathcal{P}(\mathbb{N})$.

This follows from Problems 50 and 51 and the Cantor-Schroeder-Bernstein Theorem.

Let A, B, C, and D be sets. Show that each of the following is true.

53. $\preccurlyeq$ is transitive.

Suppose that $A \preccurlyeq B$ and $B \preccurlyeq C$. Then there are functions $f: A \hookrightarrow B$ and $g: B \hookrightarrow C$. By Composition Fact 1, $g \circ f: A \hookrightarrow C$. So, $A \preccurlyeq C$. Therefore, $\preccurlyeq$ is transitive.

54. $\prec$ is transitive.

Suppose that $A \prec B$ and $B \prec C$. Then $A \preccurlyeq B$ and $B \preccurlyeq C$. By Problem 53, $A \preccurlyeq C$. Assume toward contradiction that $A \sim C$. Since $\sim$ is symmetric, $C \sim A$. In particular, $C \preccurlyeq A$. Since $C \preccurlyeq A$ and $A \preccurlyeq B$, by Problem 53, $C \preccurlyeq B$. Since $B \preccurlyeq C$ and $C \preccurlyeq B$, by the Cantor-Schroeder-Bernstein Theorem, $B \sim C$, contradicting $B \prec C$. It follows that $A \nsim C$, and thus, $A \prec C$.

55. If $A \preccurlyeq B$ and $B \prec C$, then $A \prec C$.

Suppose that $A \preccurlyeq B$ and $B \prec C$. Then $B \preccurlyeq C$. By Problem 53, $A \preccurlyeq C$. Assume toward contradiction that $A \sim C$. The rest of the argument is the same as the argument given in Problem 54.

56. If $A \prec B$ and $B \preccurlyeq C$, then $A \prec C$.

Suppose that $A \prec B$ and $B \preccurlyeq C$. Then $A \preccurlyeq B$. By Problem 53, $A \preccurlyeq C$. Assume toward contradiction that $A \sim C$. Since $\sim$ is symmetric, $C \sim A$. In particular, $C \preccurlyeq A$. Since $B \preccurlyeq C$ and $C \preccurlyeq A$, by Problem 53, $B \preccurlyeq A$. Since $A \preccurlyeq B$ and $B \preccurlyeq A$, by the Cantor-Schroeder-Bernstein Theorem, $A \sim B$, contradicting $A \prec B$. It follows that $A \nsim C$, and thus, $A \prec C$.

57. If $A \sim B$, then $\mathcal{P}(A) \sim \mathcal{P}(B)$.

Suppose that $A \sim B$. Then there exists a bijection $h: A \to B$. Define $F: \mathcal{P}(A) \to \mathcal{P}(B)$ by $F(X) = \{h(a) \mid a \in X\}$ for each $X \in \mathcal{P}(A)$.

Suppose $X, Y \in \mathcal{P}(A)$ with $F(X) = F(Y)$. Let $a \in X$. Then $h(a) \in F(X)$. Since $F(X) = F(Y)$, $h(a) \in F(Y)$. So, there is $b \in Y$ such that $h(a) = h(b)$. Since h is injective, $a = b$. So, $a \in Y$. Since $a \in X$ was arbitrary, $X \subseteq Y$. By a symmetrical argument, $Y \subseteq X$. Therefore, $X = Y$. Since $X, Y \in \mathcal{P}(A)$ were arbitrary, F is injective.

Let $Y \in \mathcal{P}(B)$ and let $X = \{a \in A \mid h(a) \in Y\}$. Then $b \in F(X)$ if and only if $b = h(a)$ for some $a \in X$ if and only if $b \in Y$ (because h is surjective). So, $F(X) = Y$. Since $Y \in \mathcal{P}(B)$ was arbitrary, F is surjective.

Since F is injective and surjective, $\mathcal{P}(A) \sim \mathcal{P}(B)$.

58. If $A \sim B$ and $C \sim D$, then $A \times C \sim B \times D$.

Suppose that $A \sim B$ and $C \sim D$. Then there exist bijections $h: A \to B$ and $k: C \to D$. Define $f: A \times C \to B \times D$ by $f(a, c) = \big(h(a), k(c)\big)$.

Suppose $(a, c), (a', c') \in A \times C$ with $f\big((a, c)\big) = f\big((a', c')\big)$. Then $\big(h(a), k(c)\big) = \big(h(a'), k(c')\big)$. So, $h(a) = h(a')$ and $k(c) = k(c')$. Since h is an injection, $a = a'$. Since k is an injection, $c = c'$. Since $a = a'$ and $c = c'$, $(a, c) = (a', c')$. Since $(a, c), (a', c') \in A \times C$ were arbitrary, f is an injection.

Now, let $(b, d) \in B \times D$. Since h and k are bijections, h^{-1} and k^{-1} exist. Let $a = h^{-1}(b)$, $c = k^{-1}(d)$. Then $f(a, c) = \big(h(a), k(c)\big) = \Big(h\big(h^{-1}(b)\big), k\big(k^{-1}(d)\big)\Big) = (b, d)$. Since $(b, d) \in B \times D$ was arbitrary, f is a surjection.

Since f is both an injection and a surjection, $A \times C \sim B \times D$.

LEVEL 5

Define $\mathcal{P}_k(\mathbb{N})$ for each $k \in \mathbb{N}$ by $\mathcal{P}_0(\mathbb{N}) = \mathbb{N}$ and $\mathcal{P}_{k+1}(\mathbb{N}) = \mathcal{P}\big(\mathcal{P}_k(\mathbb{N})\big)$ for $k > 0$.

59. Explain why $\mathcal{P}_k(\mathbb{N}) \prec \mathcal{P}_{k+1}(\mathbb{N})$ for each $k \in \mathbb{N}$.

This follows from Cantor's Theorem (Equinumerosity Fact 1).

60. Find a set B such that for all $k \in \mathbb{N}$, $\mathcal{P}_k(\mathbb{N}) \prec B$.

$\boldsymbol{B} = \bigcup\{\boldsymbol{\mathcal{P}_n}(\mathbb{N}) \mid \boldsymbol{n} \in \mathbb{N}\}$

To see that $\mathcal{P}_k(\mathbb{N}) \prec B$ for all $k \in \mathbb{N}$, begin by letting $k \in \mathbb{N}$. Since $\mathcal{P}_k(\mathbb{N}) \subseteq B$, $\mathcal{P}_k(\mathbb{N}) \preccurlyeq B$. Since k was arbitrary, we have $\mathcal{P}_k(\mathbb{N}) \preccurlyeq B$ for all $k \in \mathbb{N}$.

Again, let $k \in \mathbb{N}$. We have $\mathcal{P}_k(\mathbb{N}) \prec \mathcal{P}_{k+1}(\mathbb{N})$ and $\mathcal{P}_{k+1}(\mathbb{N}) \preccurlyeq B$. By Problem 56, $\mathcal{P}_k(\mathbb{N}) \prec B$. Since $k \in \mathbb{N}$ was arbitrary, we have shown that for all $k \in \mathbb{N}$, $\mathcal{P}_k(\mathbb{N}) \prec B$.

61. Find a set C such that $B \prec C$.

$\boldsymbol{C = \mathcal{P}(B)}$

62. Is there a set X such that $A \preccurlyeq X$ for all sets A?

No. Given any set X, $D = \mathcal{P}(X)$ is a set such that $D \npreceq X$.

Let A, B, C, and D be sets. Show that each of the following is true.

63. If $A \sim B$ and $C \sim D$, then ${}^{A}C \sim {}^{B}D$.

Suppose that $A \sim B$ and $C \sim D$. Then there exist bijections $h: A \to B$ and $k: C \to D$. Define $F: {}^{A}C \to {}^{B}D$ by $F(f)(b) = k\left(f\left(h^{-1}(b)\right)\right)$.

Suppose $f, g \in {}^{A}C$ with $F(f) = F(g)$. Let $a \in A$ and let $b = h(a)$. We have $F(f)(b) = F(g)(b)$, or equivalently, $k\left(f\left(h^{-1}(b)\right)\right) = k\left(g\left(h^{-1}(b)\right)\right)$. Since k is injective, $f\left(h^{-1}(b)\right) = g\left(h^{-1}(b)\right)$. Since $b = h(a)$, $a = h^{-1}(b)$. So, $f(a) = g(a)$. Since $a \in A$ was arbitrary, $f = g$. Since $f, g \in {}^{A}C$ were arbitrary, F is injective.

Now, let $g \in {}^{B}D$ and let's define $f \in {}^{A}C$ by $f(a) = k^{-1}\left(g\left(h(a)\right)\right)$. Let $b \in B$. Then we have $F(f)(b) = k\left(f\left(h^{-1}(b)\right)\right) = k\left(k^{-1}\left(g\left(h\left(h^{-1}(b)\right)\right)\right)\right) = g(b)$. Since $b \in B$ was arbitrary, we have $F(f) = g$. Since $g \in {}^{B}D$ was arbitrary, F is surjective.

Since F is injective and surjective, ${}^{A}C \sim {}^{B}D$.

64. ${}^{B \times C}A \sim {}^{C}({}^{B}A)$.

Let A, B, and C be sets, and define $F: {}^{B \times C}A \to {}^{C}({}^{B}A)$ by $F(f)(c)(b) = f(b, c)$.

Suppose $f, g \in {}^{B \times C}A$ with $F(f) = F(g)$. Let $c \in C$. Since $F(f) = F(g)$, $F(f)(c) = F(g)(c)$. So, for all $b \in B$, $F(f)(c)(b) = F(g)(c)(b)$. So, for all $b \in B$, $f(b, c) = g(b, c)$. Since $c \in C$ was arbitrary, for all $b \in B$ and $c \in C$, $f(b, c) = g(b, c)$. Therefore, $f = g$. Since $f, g \in {}^{B \times C}A$ were arbitrary, F is injective.

Let $k \in {}^{C}({}^{B}A)$ and define $f \in {}^{B \times C}A$ by $f(b, c) = k(c)(b)$. Then $F(f)(c)(b) = f(b, c) = k(c)(b)$. So, $F(f) = k$. Since $k \in {}^{C}({}^{B}A)$ was arbitrary, F is surjective.

Since F is injective and surjective, ${}^{B \times C}A \sim {}^{C}({}^{B}A)$.

Let $\{A_n \mid n \in \mathbb{N}\}$ be a pairwise disjoint collection of countable sets and let $\boldsymbol{P} = \{P_n \mid n \in \mathbb{N}\}$ be a partition of $\mathbb{N}$ such that each P_n is infinite (this exists by Problem 80 in Problem Set 4).

65. Explain why there is an injection $f_n: A_n \to P_n$ for each $n \in \mathbb{N}$.

A_n is countable and P_n is infinite.

Note: Since each P_n is countable, we can actually insist that each f_n is a bijection.

66. If $f: \bigcup\{A_n | n \in \mathbb{N}\} \to \mathbb{N}$ is defined by $f(x) = f_n(x)$ if $x \in A_n$, explain why f is well-defined (where f_n was described in Problem 65)

$\{A_n \mid n \in \mathbb{N}\}$ is pairwise disjoint.

67. Show that f (as defined in Problem 66) is injective.

Suppose that $x, y \in \bigcup\{A_n | n \in \mathbb{N}\}$ with $f(x) = f(y)$. There exist $n, m \in \mathbb{N}$ such that $x \in A_n$ and $y \in A_m$. So, $f(x) = f_n(x) \in P_n$ and $f(y) = f_m(y) \in P_m$. Since $f(x) = f(y)$, we have $f_n(x) = f_m(y)$. Since for $n \neq m$, $P_n \cap P_m = \emptyset$, we must have $n = m$. So, we have $f_n(x) = f_n(y)$. Since f_n is injective, $x = y$. Since $x, y \in \bigcup\{A_n | n \in \mathbb{N}\}$ were arbitrary, f is an injective function.

68. Show that a countable union of countable sets is countable.

For each $n \in \mathbb{N}$, let A_n be a countable set. By replacing each A_n by $A_n \times \{n\}$, we can assume that $\{A_n \mid n \in \mathbb{N}\}$ is a pairwise disjoint collection of sets ($A_n \sim A_n \times \{n\}$ via the bijection f sending x to (x, n)).

By Problem 80 in Problem Set 4, there is a partition $\boldsymbol{P}$ of $\mathbb{N}$ such that $\boldsymbol{P} \sim \mathbb{N}$ and for each $X \in \boldsymbol{P}$, $X \sim \mathbb{N}$, say $\boldsymbol{P} = \{P_n \mid n \in \mathbb{N}\}$.

By Problem 65, there are injections $f_n: A_n \to P_n$.

Define $f: \bigcup\{A_n | n \in \mathbb{N}\} \to \mathbb{N}$ by $f(x) = f_n(x)$ if $x \in A_n$. By Problem 66, f is well-defined. By Problem 67, f is injective.

Therefore, $\bigcup\{A_n | n \in \mathbb{N}\}$ is countable.

Problem Set 7

LEVEL 1

Determine whether each of the following sentences is an atomic statement, a compound statement, or not a statement at all:

1. Henry does not know where he is going.

This is a **compound statement**. It has the form $\neg p$, where p is the statement "Henry knows where he is going."

2. What are you doing?

This is **not a statement**. It is a question.

3. Stop doing that!

This is **not a statement**. It is a command.

4. $x \neq 53$.

This is **not a statement**. It has an unknown variable.

5. I like the song *Crimson and Clover*.

This is an **atomic statement**. Even though the word "and" appears in the statement, here it is part of the name of the song. It is not being used as a connective.

6. If lobsters are birds, then we have a problem.

This is a **compound statement**. It has the form $p \to q$, where p is the statement "Lobsters are birds," and q is the statement "We have a problem."

7. $5 = 6$ or $5 \neq 6$.

This is a **compound statement**. It has the form $p \vee \neg p$, where p is the statement "$5 = 6$."

8. This sentence has five words.

This is **not a statement** because it is self-referential. Self-referential sentences can cause problems. For example, observe that the negation of this sentence would be "This sentence does not have five words." The sentence and its negation both appear to be true. That would be a problem. It's a good thing they're not statements!

9. A quadrilateral is a rhombus if and only if the diagonals of the quadrilateral are perpendicular.

This is a **compound statement**. It has the form $p \leftrightarrow q$, where p is the statement "A quadrilateral is a rhombus," and q is the statement "The diagonals of the quadrilateral are perpendicular."

10. I cannot speak Japanese, but I can speak Italian.

This is a **compound statement**. It has the form $\neg p \wedge q$, where p is the statement "I can speak Japanese," and q is the statement "I can speak Italian." Note that in sentential logic, the word "but" has the same meaning as the word "and." In English, the word "but" is used to introduce contrast with the part of the sentence that has already been mentioned. However, logically it is no different from the word "and."

What is the negation of each of the following statements?

11. A tomato is a fruit.

A tomato is not a fruit.

12. Cats have nine lives.

Cats do not have nine lives.

13. $76 \geq 72$.

$\mathbf{76 < 72}$

Note: In words, $76 \geq 72$ can be read, "Seventy six is greater than or equal to seventy two."

The negation of this statement is "Seventy six is not greater than or equal to seventy two" (symbolically, we can write $76 \ngeq 72$). Using a basic property of the number systems we are most familiar with, $76 \ngeq 72$ is equivalent to $76 < 72$.

It follows that the negation of $76 \geq 72$ is equivalent to $76 < 72$.

14. You will not believe what I just heard.

You will believe what I just heard.

Note: Technically, the negation is "It is not the case that you will not believe what I just heard." However, by the law of double negation, this is logically equivalent to "You will believe what I just heard."

15. The function f has a discontinuity at $x = 3$.

The function $\boldsymbol{f}$ does not have a discontinuity at $\boldsymbol{x = 3}$.

16. The fundamental group of the circle is isomorphic to the set of integers.

The fundamental group of the circle is not isomorphic to the set of integers.

17. Some birds can fly.

No birds can fly.

Note: "Some birds can fly" has the form $\exists x(P(x))$. The negation of this statement is $\neg\exists x(P(x))$, which is logically equivalent to $\forall x(\neg P(x))$. This is translated as "Every bird cannot fly," or equivalently, "No birds can fly."

18. Every summer night is warm.

Some summer nights are not warm.

Note: "Every summer night is warm" has the form $\forall x(P(x))$. The negation of this statement is $\neg\forall x(P(x))$, which is logically equivalent to $\exists x(\neg P(x))$. This is translated as "There is a summer night that is not warm," or equivalently, "Some summer nights are not warm."

LEVEL 2

Let p represent the statement "5 is an odd integer," let q represent the statement "Brazil is in Europe," and let r represent the statement "A lobster is an insect." Rewrite each of the following symbolic statements in words, and state the truth value of each statement:

Note: p has truth value T, while q and r both have truth value F (a lobster is a crustacean, not an insect).

19. $p \vee q$

$p \vee q$ represents "**5 is an odd integer or Brazil is in Europe.**" Since p has truth value T, it follows that $p \vee q$ has truth value **T**.

20. $\neg r$

$\neg r$ represents "**A lobster is not an insect.**" Since r has truth value F, it follows that $\neg r$ has truth value **T**.

21. $p \to q$

$p \to q$ represents "**If 5 is an odd integer , then Brazil is in Europe.**" Since p has truth value T and q has truth value F, it follows that $p \to q$ has truth value **F**.

22. $p \leftrightarrow r$

$p \leftrightarrow r$ represents "**5 is an odd integer if and only if a lobster is an insect.**" Since p and r have opposite truth values, $p \leftrightarrow r$ has truth value **F**.

23. $\neg q \wedge r$

$\neg q \wedge r$ represents "**Brazil is not in Europe and a lobster is an insect.**" Since r has truth value F, it follows that $\neg q \wedge r$ has truth value **F**.

24. $\neg(p \wedge q)$

$\neg(p \wedge q)$ represents "**It is not the case that 5 is an odd integer and Brazil is in Europe.**" Since $p \wedge q$ has truth value F (because q has truth value F), it follows that $\neg(p \wedge q)$ has truth value **T**.

25. $\neg p \vee \neg q$

$\neg p \vee \neg q$ represents "**5 is not an odd integer or Brazil is not in Europe.**" Since $\neg q$ has truth value T (do you see why?), it follows that $\neg p \vee \neg q$ has truth value **T**.

26. $(p \wedge q) \to r$

$(p \wedge q) \to r$ represents "**If 5 is an odd integer and Brazil is in Europe, then a lobster is an insect.**" Since $p \wedge q$ has truth value F (because q has truth value F), it follows that $(p \wedge q) \to r$ has truth value **T**.

Consider the compound sentence "You can have a cookie or ice cream." In English this would most likely mean that you can have one or the other but not both. The word "or" used here is generally called an "exclusive or" because it excludes the possibility of both. The disjunction is an "inclusive or."

27. Using the symbol $\oplus$ for exclusive or, draw the truth table for this connective.

p	q	$p \oplus q$
T	**T**	**F**
T	**F**	**T**
F	**T**	**T**
F	**F**	**F**

28. Using only the logical connectives $\neg$, $\wedge$, and $\vee$, produce a statement using the propositional variables p and q that has the same truth values as $p \oplus q$.

We want to express that p is true or q is true, but p and q are not both true. Expressed in symbols, this is $\boldsymbol{(p \vee q) \wedge \neg(p \wedge q)}$.

Note: (1) Let's check that $(p \vee q) \wedge \neg(p \wedge q)$ behaves as desired.

If p and q are both true, then $\neg(p \wedge q) \equiv \text{F}$, and so, $(p \vee q) \wedge \neg(p \wedge q) \equiv (p \vee q) \wedge \text{F} \equiv \text{F}$.

If p and q are both false, then $p \vee q \equiv \text{F}$, and so, $(p \vee q) \wedge \neg(p \wedge q) \equiv \text{F} \wedge \neg(p \wedge q) \equiv \text{F}$.

Finally, if p and q have opposite truth values, then $p \vee q \equiv \text{T}$ and $\neg(p \wedge q) \equiv \text{T}$ (because $p \wedge q \equiv \text{F}$). Therefore, $(p \vee q) \wedge \neg(p \wedge q) \equiv \text{T} \wedge \text{T} \equiv \text{T}$.

(2) Recall that the word "but" is logically the same as the word "and" (see Problem 10 above).

(3) Another way to see that $p \oplus q$ has the same truth values as $(p \vee q) \wedge \neg(p \wedge q)$ is to draw the truth tables for each and observe that row by row they have the same truth values. We do this below.

p	q	$p \oplus q$	$p \vee q$	$p \wedge q$	$\neg(p \wedge q)$	$(p \vee q) \wedge \neg(p \wedge q)$
T	T	F	T	T	F	F
T	F	T	T	F	T	T
F	T	T	T	F	T	T
F	F	F	F	F	T	F

Observe that the third column of the truth table corresponds to $p \oplus q$, the last (seventh) column corresponds to $(p \vee q) \wedge \neg(p \wedge q)$, and both these columns have the same truth values.

(4) In this problem, we showed that $p \oplus q \equiv (p \vee q) \wedge \neg(p \wedge q)$.

Determine whether each occurrence of each variable in the given formula is free or bound. Is the given formula a sentence?

29. $x \in y$

x and y are both free. The formula is **not a sentence**.

30. $\forall x(x \in y)$

x is bound and y is free. The formula is **not a sentence**.

31. $\exists y\big(x = y \rightarrow \forall x(y \in x)\big)$

The first instance of x is free and the second is bound. Both instances of y are bound. The formula is **not a sentence**.

32. $\forall x \forall y \exists z\big((x \in y \wedge x \in z) \leftrightarrow x = z\big) \vee \big(x \neq y \rightarrow \forall z(x \in z)\big)$

The first three instances of x are bound and the last two instances of x are free. The first instance of y is bound and the second instance of y is free. All three instances of z are bound. The formula is **not a sentence**.

LEVEL 3

Consider the four distinct propositional variables p, q, r, and s.

33. How many different truth assignments are there for this list of propositional variables?

$2 \cdot 2 \cdot 2 \cdot 2 = \mathbf{16}$

34. How many different truth assignments are there for this list of propositional variables such that p is true and q is false?

$2 \cdot 2 = \mathbf{4}$

35. How many different truth assignments are there for this list of propositional variables such that q, r, and s are all true?

2

36. How many different truth assignments are there for a list of 5 propositional variables?

$2 \cdot 2 \cdot 2 \cdot 2 \cdot 2 = \mathbf{32}$

Let p, q, and r represent true statements. Compute the truth value of each of the following compound statements:

37. $(p \vee q) \vee r$

Detailed solution: $(p \vee q) \vee r \equiv (\mathrm{T} \vee \mathrm{T}) \vee \mathrm{T} \equiv \mathrm{T} \vee \mathrm{T} \equiv \mathbf{T}$.

Quicker solution: $(p \vee q) \vee r \equiv (p \vee q) \vee \mathrm{T} \equiv \mathbf{T}$.

38. $(p \vee q) \wedge \neg r$

Detailed solution: $(p \vee q) \wedge \neg r \equiv (\mathrm{T} \vee \mathrm{T}) \wedge \neg \mathrm{T} \equiv \mathrm{T} \wedge \mathrm{F} \equiv \mathbf{F}$.

Quicker solution: $(p \vee q) \wedge \neg r \equiv (p \vee q) \wedge \mathrm{F} \equiv \mathbf{F}$.

39. $\neg p \rightarrow (q \vee r)$

Detailed solution: $\neg p \rightarrow (q \vee r) \equiv \neg \mathrm{T} \rightarrow (\mathrm{T} \vee \mathrm{T}) \equiv \mathrm{F} \rightarrow \mathrm{T} \equiv \mathbf{T}$.

Quicker solution: $\neg p \rightarrow (q \vee r) \equiv \mathrm{F} \rightarrow (q \vee r) \equiv \mathbf{T}$.

40. $\neg(p \leftrightarrow \neg q) \wedge r$

Detailed solution: $\neg(p \leftrightarrow \neg q) \wedge r \equiv \neg(\mathrm{T} \leftrightarrow \neg \mathrm{T}) \wedge \mathrm{T} \equiv \neg(\mathrm{T} \leftrightarrow \mathrm{F}) \wedge \mathrm{T} \equiv \neg \mathrm{F} \wedge \mathrm{T} \equiv \mathrm{T} \wedge \mathrm{T} \equiv \mathbf{T}$.

Quicker solution: $\neg(p \leftrightarrow \neg q) \wedge r \equiv \neg(\mathrm{T} \leftrightarrow \mathrm{F}) \wedge \mathrm{T} \equiv \neg \mathrm{F} \wedge \mathrm{T} \equiv \mathrm{T} \wedge \mathrm{T} \equiv \mathbf{T}$.

41. $\neg[p \wedge (\neg q \rightarrow r)]$

Detailed solution: $\neg[p \wedge (\neg q \rightarrow r)] \equiv \neg[\mathrm{T} \wedge (\neg \mathrm{T} \rightarrow \mathrm{T})] \equiv \neg[\mathrm{T} \wedge (\mathrm{F} \rightarrow \mathrm{T})] \equiv \neg[\mathrm{T} \wedge \mathrm{T}] \equiv \neg \mathrm{T} \equiv \mathbf{F}$.

Quicker solution: $\neg[p \wedge (\neg q \rightarrow r)] \equiv \neg[p \wedge (\mathrm{F} \rightarrow r)] \equiv \neg[p \wedge \mathrm{T}] \equiv \neg[\mathrm{T} \wedge \mathrm{T}] \equiv \neg \mathrm{T} \equiv \mathbf{F}$.

42. $\neg[(\neg p \vee \neg q) \leftrightarrow \neg r]$

Detailed solution:

$$\neg[(\neg p \vee \neg q) \leftrightarrow \neg r] \equiv \neg[(\neg \mathrm{T} \vee \neg \mathrm{T}) \leftrightarrow \neg \mathrm{T}] \equiv \neg[(\mathrm{F} \vee \mathrm{F}) \leftrightarrow \mathrm{F}] \equiv \neg[\mathrm{F} \leftrightarrow \mathrm{F}] \equiv \neg \mathrm{T} \equiv \mathbf{F}.$$

Quicker solution: $\neg[(\neg p \vee \neg q) \leftrightarrow \neg r] \equiv \neg[\mathrm{F} \leftrightarrow \mathrm{F}] \equiv \neg\mathrm{T} \equiv \mathbf{F}$.

43. $p \to (q \to \neg r)$

Detailed solution: $p \to (q \to \neg r) \equiv \mathrm{T} \to (\mathrm{T} \to \neg\mathrm{T}) \equiv \mathrm{T} \to (\mathrm{T} \to \mathrm{F}) \equiv \mathrm{T} \to \mathrm{F} \equiv \mathbf{F}$.

Quicker solution: $p \to (q \to \neg r) \equiv \mathrm{T} \to (\mathrm{T} \to \mathrm{F}) \equiv \mathrm{T} \to \mathrm{F} \equiv \mathbf{F}$.

44. $\neg[\neg p \to (q \to \neg r)]$

Detailed solution:

$$\neg[\neg p \to (q \to \neg r)] \equiv \neg[\neg\mathrm{T} \to (\mathrm{T} \to \neg\mathrm{T})] \equiv \neg[\mathrm{F} \to (\mathrm{T} \to \mathrm{F})] \equiv \neg[\mathrm{F} \to \mathrm{F}] \equiv \neg\mathrm{T} \equiv \mathbf{F}.$$

Quicker solution: $\neg[\neg p \to (q \to \neg r)] \equiv \neg[\mathrm{F} \to (q \to \neg r)] \equiv \neg\mathrm{T} \equiv \mathbf{F}$.

Determine if each of the following statements is a tautology, a contradiction, or neither.

45. $p \wedge p$

If $p \equiv \mathrm{T}$, then $p \wedge p \equiv \mathrm{T} \wedge \mathrm{T} \equiv \mathrm{T}$. If $p \equiv \mathrm{F}$, then $p \wedge p \equiv \mathrm{F} \wedge \mathrm{F} \equiv \mathrm{F}$. **Neither**

46. $p \wedge \neg p$

$p \wedge \neg p \equiv \mathrm{F}$. **Contradiction**

47. $(p \vee \neg p) \to (p \wedge \neg p)$

$(p \vee \neg p) \to (p \wedge \neg p) \equiv \mathrm{T} \to \mathrm{F} \equiv \mathrm{F}$. **Contradiction**

48. $\neg(p \vee q) \leftrightarrow (\neg p \wedge \neg q)$

Since $\neg(p \vee q) \equiv \neg p \wedge \neg q$ (De Morgan's law), $\neg(p \vee q) \leftrightarrow (\neg p \wedge \neg q)$ is a **Tautology**.

49. $p \to (\neg q \wedge r)$

If $p \equiv \mathrm{F}$, then we have $p \to (\neg q \wedge r) \equiv \mathrm{F} \to (\neg q \wedge r) \equiv \mathrm{T}$. If $p \equiv \mathrm{T}$ and $r \equiv \mathrm{F}$, then we have $p \to (\neg q \wedge r) \equiv \mathrm{T} \to (\neg q \wedge \mathrm{F}) \equiv \mathrm{T} \to \mathrm{F} \equiv \mathrm{F}$. **Neither**

50. $(p \leftrightarrow q) \to (p \to q)$

If p and q have the same truth value, then we have $p \leftrightarrow q \equiv \mathrm{T}$ and $p \to q \equiv \mathrm{T}$, and therefore, $(p \leftrightarrow q) \to (p \to q) \equiv \mathrm{T} \to \mathrm{T} \equiv \mathrm{T}$. If p and q have opposite truth values, then $p \leftrightarrow q \equiv \mathrm{F}$, and therefore, we have $(p \leftrightarrow q) \to (p \to q) \equiv \mathrm{F} \to (p \to q) \equiv \mathrm{T}$. Since all possible truth assignments of the propositional variables lead to a truth value of T, $(p \leftrightarrow q) \to (p \to q)$ is a **Tautology**.

LEVEL 4

Assume that the given compound statement is true. Determine the truth value of each propositional variable.

51. $p \wedge q$

If $p \equiv \text{F}$ or $q \equiv \text{F}$, then $p \wedge q \equiv \text{F}$. Therefore, $\boldsymbol{p} \equiv \mathbf{T}$ and $\boldsymbol{q} \equiv \mathbf{T}$.

52. $\neg(p \to q)$

Since $\neg(p \to q)$ is true, $p \to q$ is false. Therefore, $\boldsymbol{p} \equiv \mathbf{T}$ and $\boldsymbol{q} \equiv \mathbf{F}$.

53. $p \leftrightarrow [\neg(p \wedge q)]$

If $p \equiv \text{F}$, then $p \wedge q \equiv \text{F}$, and so, $p \leftrightarrow [\neg(p \wedge q)] \equiv \text{F} \leftrightarrow \text{T} \equiv \text{F}$. So, $\boldsymbol{p} \equiv \mathbf{T}$. It follows that $\neg(p \wedge q) \equiv \text{T}$, and so $p \wedge q \equiv \text{F}$. Since $p \equiv \text{T}$, we must have $\boldsymbol{q} \equiv \mathbf{F}$.

54. $[p \wedge (q \vee r)] \wedge \neg r$

As in Problem 45, we must have $p \wedge (q \vee r) \equiv \text{T}$ and $\neg r \equiv \text{T}$. So, $\boldsymbol{p} \equiv \mathbf{T}$, $q \vee r \equiv \text{T}$, and $\boldsymbol{r} \equiv \mathbf{F}$. Since $q \vee r \equiv \text{T}$ and $r \equiv \text{F}$, we must have $\boldsymbol{q} \equiv \mathbf{T}$.

Let p represent a true statement. Decide if this is enough information to determine the truth value of each of the following statements. If so, state that truth value.

55. $p \vee q$

$(p \vee q) \equiv \text{T} \vee q \equiv \mathbf{T}$.

56. $p \to q$

$p \to q \equiv \text{T} \to q$. If $q \equiv \text{T}$, we get $\text{T} \to \text{T} \equiv \text{T}$. If $q \equiv \text{F}$, we get $\text{T} \to \text{F} \equiv \text{F}$. **There is not enough information**.

57. $\neg p \to \neg(q \vee \neg r)$

$\neg p \to \neg(q \vee \neg r) \equiv \text{F} \to \neg(q \vee \neg r) \equiv \mathbf{T}$.

58. $\neg(\neg p \wedge q) \leftrightarrow p$

$\neg(\neg p \wedge q) \leftrightarrow p \equiv \neg(\text{F} \wedge q) \leftrightarrow \text{T} \equiv \neg\text{F} \leftrightarrow \text{T} \equiv \text{T} \leftrightarrow \text{T} \equiv \mathbf{T}$.

59. $(p \leftrightarrow q) \leftrightarrow \neg p$

$(p \leftrightarrow q) \leftrightarrow \neg p \equiv (\text{T} \leftrightarrow q) \leftrightarrow \text{F}$. If $q \equiv \text{T}$, we get $(\text{T} \leftrightarrow \text{T}) \leftrightarrow \text{F} \equiv \text{T} \leftrightarrow \text{F} \equiv \text{F}$. If $q \equiv \text{F}$, we get $(\text{T} \leftrightarrow \text{F}) \leftrightarrow \text{F} \equiv \text{F} \leftrightarrow \text{F} \equiv \text{T}$. **There is not enough information**.

60. $\neg[(\neg p \wedge \neg q) \leftrightarrow \neg r]$

$\neg[(\neg p \wedge \neg q) \leftrightarrow \neg r] \equiv \neg[(\mathrm{F} \wedge \neg q) \leftrightarrow \neg r] \equiv \neg(\mathrm{F} \leftrightarrow \neg r)$. If $r \equiv \mathrm{T}$, we get $\neg\mathrm{T} \equiv \mathrm{F}$. If $r \equiv \mathrm{F}$, we get $\neg\mathrm{F} \equiv \mathrm{T}$. **There is not enough information**.

61. $[(p \wedge \neg p) \rightarrow p] \wedge (p \vee \neg p)$

$[(p \wedge \neg p) \rightarrow p] \wedge (p \vee \neg p) \equiv [(\mathrm{T} \wedge \mathrm{F}) \rightarrow \mathrm{T}] \wedge (\mathrm{T} \vee \mathrm{F}) \equiv [\mathrm{F} \rightarrow \mathrm{T}] \wedge \mathrm{T} \equiv \mathrm{T} \wedge \mathrm{T} \equiv \mathbf{T}$.

62. $r \rightarrow [\neg q \rightarrow (\neg p \rightarrow \neg r)]$

$r \rightarrow [\neg q \rightarrow (\neg p \rightarrow \neg r)] \equiv r \rightarrow [\neg q \rightarrow (\mathrm{F} \rightarrow \neg r)] \equiv r \rightarrow [\neg q \rightarrow \mathrm{T}] \equiv r \rightarrow \mathrm{T} \equiv \mathbf{T}$.

For each of the following pairs of statements A and B, show that $A \equiv B$.

63. $A = p \wedge q,\ B = q \wedge p$

Draw the truth tables.

64. $A = (p \vee q) \vee r,\ B = p \vee (q \vee r)$

Draw truth tables.

65. $A = p \wedge (q \vee r),\ B = (p \wedge q) \vee (p \wedge r)$

Draw truth tables.

66. $A = (p \vee q) \wedge p,\ B = p$

If $p \equiv \mathrm{T}$, then $p \vee q \equiv \mathrm{T} \vee q \equiv \mathrm{T}$. So, $A \equiv \mathrm{T} \wedge \mathrm{T} \equiv \mathrm{T}$. Also, $B \equiv \mathrm{T}$. If $p \equiv \mathrm{F}$, then $A \equiv (p \vee q) \wedge \mathrm{F} \equiv \mathrm{F}$ and $B \equiv \mathrm{F}$. So, all four possible truth assignments of p and q lead to the same truth value for ϕ and ψ. It follows that $\phi \equiv \psi$.

Note: We have just proved the first **absorption law**.

67. $A = p \leftrightarrow q,\ B = (p \rightarrow q) \wedge (q \rightarrow p)$

If $p \equiv \mathrm{T}$, then we have $A \equiv \mathrm{T} \leftrightarrow q \equiv q$, $B \equiv (\mathrm{T} \rightarrow q) \wedge (q \rightarrow \mathrm{T}) \equiv q \wedge \mathrm{T} \equiv q$. If $p \equiv \mathrm{F}$, then we have $A \equiv \mathrm{F} \leftrightarrow q \equiv \neg q$, $B \equiv (\mathrm{F} \rightarrow q) \wedge (q \rightarrow \mathrm{F}) \equiv \mathrm{T} \wedge \neg q \equiv \neg q$. So, all four possible truth assignments of p and q lead to the same truth value for A and B. It follows that $A \equiv B$.

Note: We have just proved the **law of the biconditional**.

68. $A = \neg(p \rightarrow q),\ B = p \wedge \neg q$

If $p \equiv \mathrm{F}$, then we have $A \equiv \neg(\mathrm{F} \rightarrow q) \equiv \neg\mathrm{T} \equiv \mathrm{F}$, $B \equiv \mathrm{F} \wedge \neg q \equiv \mathrm{F}$. If $q \equiv \mathrm{T}$, then we have $A \equiv \neg(p \rightarrow \mathrm{T}) \equiv \neg\mathrm{T} \equiv \mathrm{F}$, $B \equiv p \wedge \neg\mathrm{T} \equiv p \wedge \mathrm{F} \equiv \mathrm{F}$. Finally, if $p \equiv \mathrm{T}$ and $q \equiv \mathrm{F}$, then we have $A \equiv \neg(\mathrm{T} \rightarrow \mathrm{F}) \equiv \neg\mathrm{F} \equiv \mathrm{T}$, $B \equiv \mathrm{T} \wedge \neg\mathrm{F} \equiv \mathrm{T} \wedge \mathrm{T} \equiv \mathrm{T}$. So, all four possible truth assignments of p and q lead to the same truth value for A and B. It follows that $A \equiv B$.

Simplify each statement.

69. $p \vee (p \wedge \neg p)$

$p \vee (p \wedge \neg p) \equiv p \vee \mathrm{F} \equiv \boldsymbol{p}$.

70. $(p \wedge q) \vee \neg p$

$(p \wedge q) \vee \neg p \equiv (p \vee \neg p) \wedge (q \vee \neg p) \equiv \mathrm{T} \wedge (q \vee \neg p) \equiv q \vee \neg p \equiv \boldsymbol{\neg p \vee q}$.

71. $\neg p \to (\neg q \to p)$

$\neg p \to (\neg q \to p) \equiv p \vee (\neg q \to p) \equiv p \vee (q \vee p) \equiv p \vee (p \vee q) \equiv (p \vee p) \vee q \equiv \boldsymbol{p \vee q}$.

72. $(p \wedge \neg q) \vee p$

$(p \wedge \neg q) \vee p \equiv \boldsymbol{p}$ (Absorption).

73. $[(q \wedge p) \vee q] \wedge [(q \vee p) \wedge p]$

$[(q \wedge p) \vee q] \wedge [(q \vee p) \wedge p] \equiv [(q \wedge p) \vee q] \wedge [(p \vee q) \wedge p] \equiv q \wedge p$ (Absorption) $\equiv \boldsymbol{p \wedge q}$.

LEVEL 5

Without drawing a truth table or using List 7.28, show that each of the following is a tautology.

74. $[p \wedge (q \vee r)] \leftrightarrow [(p \wedge q) \vee (p \wedge r)]$

If $p \equiv \mathrm{F}$, then $p \wedge (q \vee r) \equiv \mathrm{F}$, $p \wedge q \equiv \mathrm{F}$, and $p \wedge r \equiv \mathrm{F}$. So, $(p \wedge q) \vee (p \wedge r) \equiv \mathrm{F}$. It follows that $[p \wedge (q \vee r)] \leftrightarrow [(p \wedge q) \vee (p \wedge r)] \equiv \mathrm{F} \leftrightarrow \mathrm{F} \equiv \mathbf{T}$.

If $p \equiv \mathrm{T}$ and $q \equiv \mathrm{T}$, then $p \wedge (q \vee r) \equiv \mathrm{T} \wedge \mathrm{T} \equiv \mathrm{T}$ and $(p \wedge q) \vee (p \wedge r) \equiv \mathrm{T} \vee (p \wedge r) \equiv \mathrm{T}$. It follows that $[p \wedge (q \vee r)] \leftrightarrow [(p \wedge q) \vee (p \wedge r)] \equiv \mathrm{T} \leftrightarrow \mathrm{T} \equiv \mathbf{T}$.

If $p \equiv \mathrm{T}$ and $q \equiv \mathrm{F}$, then $p \wedge (q \vee r) \equiv \mathrm{T} \wedge r \equiv r$ and $(p \wedge q) \vee (p \wedge r) \equiv \mathrm{F} \vee r \equiv r$. It follows that $[p \wedge (q \vee r)] \leftrightarrow [(p \wedge q) \vee (p \wedge r)] \equiv r \leftrightarrow r \equiv \mathbf{T}$.

Notes: (1) We can display this reasoning visually as follows:

$$
\begin{array}{l}
[p \wedge (q \vee r)] \leftrightarrow [(p \wedge q) \vee (p \wedge r)] \\
\mathrm{F}\ \ \mathrm{F} \qquad\qquad\ \ \mathbf{T} \quad \mathrm{F}\ \ \mathrm{F} \quad\ \ \mathrm{F}\ \ \mathrm{F}\,\mathrm{F} \\
\mathrm{T}\ \ \mathrm{T}\ \ \mathrm{T}\,\mathrm{T} \quad\ \ \mathbf{T} \quad \mathrm{T}\ \ \mathrm{T}\,\mathrm{T}\ \ \mathrm{T} \\
\mathrm{T}\ \ r\ \ \ \mathrm{F}\,r \quad\ \ \ \mathbf{T} \quad \mathrm{T}\ \ \mathrm{F}\,\mathrm{F} \ \ r\ \ \mathrm{T}\,r
\end{array}
$$

Each row of truth values is placed in the order suggested by the solution above. For example, for the first row, we start by writing F under each p because we are assuming that $p \equiv \mathrm{F}$. Next, since the conjunction of F with anything else is F, we write F under each $\wedge$ (there are three that appear). Next, since $\mathrm{F} \vee \mathrm{F} \equiv \mathrm{F}$, we write F under the rightmost $\vee$. Finally, since $\mathrm{F} \leftrightarrow \mathrm{F} \equiv \mathrm{T}$, we write T under $\leftrightarrow$. This is the truth value of the entire statement, and therefore, we are done with the case $p \equiv \mathrm{F}$. The other two rows work the same way.

(2) We used two of the identity laws in the third part of the solution: $\mathrm{T} \wedge r \equiv r$ and $\mathrm{F} \vee r \equiv r$.

75. $\big[[(p \wedge q) \to r] \to s\big] \to [(p \to r) \to s]$

If $s \equiv \mathrm{T}$, then $(p \to r) \to s \equiv \mathrm{T}$, and therefore, $\big[[(p \wedge q) \to r] \to s\big] \to [(p \to r) \to s] \equiv \mathbf{T}$.

Now, assume $s \equiv \mathrm{F}$, and either $p \equiv \mathrm{F}$ or $q \equiv \mathrm{F}$. Then $p \wedge q \equiv \mathrm{F}$, and so $(p \wedge q) \to r \equiv \mathrm{T}$. Therefore, $[(p \wedge q) \to r] \to s \equiv \mathrm{F}$, and so, $\big[[(p \wedge q) \to r] \to s\big] \to [(p \to r) \to s] \equiv \mathbf{T}$.

Finally, assume $s \equiv \mathrm{F}$, $p \equiv \mathrm{T}$, and $q \equiv \mathrm{T}$. Then $p \wedge q \equiv \mathrm{T}$, and so, $(p \wedge q) \to r \equiv r$. Therefore, $[(p \wedge q) \to r] \to s \equiv \neg r$. Also, $p \to r \equiv r$, and so $(p \to r) \to s \equiv \neg r$. So, we get $\neg r \to \neg r \equiv \mathbf{T}$.

Note: The dedicated reader should display this reasoning visually, as was done in Note 1 following the solution to Problem 74 above.

Let n be a positive integer (in other words, n is one of the numbers $1, 2, 3, 4, \ldots$) and let A be a statement involving n propositional variables. Determine how many rows are in the truth table for A if n is equal to each of the following:

76. $n = 6$

$2 \cdot 2 \cdot 2 \cdot 2 \cdot 2 \cdot 2 = 2^6 = \mathbf{64}$

77. $n = 10$

$2 \cdot 2 \cdot 2 \cdot 2 \cdot 2 \cdot 2 \cdot 2 \cdot 2 \cdot 2 \cdot 2 = 2^{10} = \mathbf{1024}$

78. n is an arbitrary positive integer (provide an explicit expression involving n.)

$\mathbf{2^n}$

Use the axioms of ZF to show each of the following:

79. If x, y, and z are sets, then the ordered triple (x, y, z) is a set.

Since x and y are sets, by Exercise 7.46, $w = (x, y)$ is a set. Since w and z are sets, again by Exercise 7.46, (w, z) is a set. Therefore, $(x, y, z) = \big((x, y), z\big) = (w, z)$ is a set.

80. If $x_1, x_2, \ldots, x_{n+1}$ are sets and every n-tuple of sets is a set, then $(x_1, x_2, \ldots, x_n, x_{n+1})$ is a set.

Let $x = (x_1, x_2, \dots x_n)$. By assumption, x is a set. Since x and x_{n+1} are sets, by Exercise 7.46, (x, x_{n+1}) is a set. Therefore, $(x_1, x_2, \dots, x_n, x_{n+1}) = \big((x_1, x_2, \dots, x_n), x_{n+1}\big) = (x, x_{n+1})$ is a set.

81. If A and B are sets, then $\mathcal{P}\big(\mathcal{P}(A \cup B)\big)$ is a set.

By the pairing axiom, $\{A, B\}$ is a set. So, by the union axiom, $A \cup B = \bigcup\{A, B\}$ is a set. By applying the power set axiom twice, we see that $\mathcal{P}\big(\mathcal{P}(A \cup B)\big)$ is a set.

82. If A and B are sets, then $A \times B$ is a set.

If $a \in A$ and $b \in B$, then $\{a\}$ and $\{a, b\}$ are in $\mathcal{P}(A \cup B)$. So, $(a, b) = \big\{\{a\}, \{a, b\}\big\} \in \mathcal{P}\big(\mathcal{P}(A \cup B)\big)$. By Problem 81, $\mathcal{P}\big(\mathcal{P}(A \cup B)\big)$ is a set. $A \times B = \big\{x \in \mathcal{P}\big(\mathcal{P}(A \cup B)\big) \,\big|\, x = (a, b) \wedge a \in A \wedge b \in B\big\}$ is now seen to be a set by Bounded Comprehension.

Problem Set 8

LEVEL 1

Determine if each of the following ordered sets is well-ordered. Note that $<$ is always the usual order on the given set.

1. $(\mathbb{Z}^+, <)$

Well-ordered

2. $((0,1), <)$

Not well-ordered ($(0,1)$ has no least element)

3. $([1,8), <)$

Not well-ordered ($(1,8) \subseteq [1,8)$ and $(1,8)$ has no least element)

4. $([1,8) \cap \mathbb{Q}, <)$

Not well-ordered ($(1,8) \cap \mathbb{Q} \subseteq [1,8)$ and $(1,8) \cap \mathbb{Q}$ has no least element)

5. $(\{0,1,2,3,4\}, <)$

Well-ordered

6. $(2\mathbb{N}, <)$

Well-ordered

7. $(X, <_D)$, where $X = \{x \mid x \text{ is a word in the English language}\}$ and $<_D$ is the dictionary order on X.

Well-ordered

Find an ordinal isomorphic to the given well-ordered set.

8. $(2\mathbb{N} + 1, <)$

$\boldsymbol{\omega}$

9. $(\{a, b, c, d, e, f, g\}, <)$, where $<$ orders the given set alphabetically

6

10. $(A, <^*)$ (from Exercise 8.2)

$\boldsymbol{\omega + 1}$

11. $(\mathbb{Z}^-, >)$

$\boldsymbol{\omega}$

12. $\text{pred}(\mathbb{N}, 0)$

0

13. $\text{pred}(\mathbb{N}, 30)$

30

14. $\text{pred}(A, \star)$, where $(A, <^\star)$ is as defined in Exercise 8.2

$\boldsymbol{\omega}$

Use cardinal arithmetic to compute each of the following:

15. $5 + 11$

16

16. $7 + \omega$

$\boldsymbol{\omega}$

17. $\omega + 7$

$\boldsymbol{\omega}$

18. $\omega + \omega$

$\boldsymbol{\omega}$

19. $\omega \cdot \omega$

$\boldsymbol{\omega}$

20. $10 \cdot \omega$

$\boldsymbol{\omega}$

LEVEL 2

Use ordinal arithmetic to write each of the following ordinals in the form $\omega \cdot a + b$, where $a, b \in \omega$.

21. $7 + 3$

$\mathbf{10 = \omega \cdot 0 + 10}$

22. $7 + \omega$

$\boldsymbol{\omega = \omega \cdot 1 + 0}$

23. $\omega + 7$

$\boldsymbol{\omega + 7 = \omega \cdot 1 + 7}$

24. 7ω

$\boldsymbol{\omega = \omega \cdot 1 + 0}$

25. $\omega \cdot 7$

$\boldsymbol{\omega \cdot 7 = \omega \cdot 7 + 0}$

26. $7\omega + 7$

$\boldsymbol{\omega + 7 = \omega \cdot 1 + 7}$

27. $\omega \cdot 7 + 7$

$\boldsymbol{\omega \cdot 7 + 7}$

28. $7 + 7\omega$

$7 + \omega = \boldsymbol{\omega = \omega \cdot 1 + 0}$

29. $7 + \omega \cdot 7$

$\boldsymbol{\omega \cdot 7 = \omega \cdot 7 + 0}$

30. $7\omega + \omega$

$\omega + \omega = \boldsymbol{\omega \cdot 2 = \omega \cdot 2 + 0}$

31. $7\omega + 7\omega$

$\omega + \omega = \boldsymbol{\omega \cdot 2 = \omega \cdot 2 + 0}$

32. $\omega \cdot 7 + \omega$

$\boldsymbol{\omega \cdot 8 = \omega \cdot 8 + 0}$

33. $\omega + \omega \cdot 7$

$\boldsymbol{\omega \cdot 8 = \omega \cdot 8 + 0}$

34. $\omega \cdot 7 + \omega \cdot 7$

$\boldsymbol{\omega \cdot 14 = \omega \cdot 14 + 0}$

Determine if each of the following statements is true or false. If true, explain why. If false, provide a counterexample.

35. Every well-ordered set is isomorphic to an initial segment of the set of natural numbers.

False: For example, $\omega + 1$ is **not** isomorphic to an initial segment of the natural numbers.

36. If $(A, <)$ is an infinite well-ordered set, then ω is isomorphic to an initial segment of $(A, <)$.

True: This follows from Well-ordered Set Fact 2, together with the fact that ω is the least infinite ordinal.

37. Every transitive set is an ordinal.

False: For example, $\mathcal{P}(\{0,1\}) = \mathcal{P}(\{\emptyset, \{\emptyset\}\}) = \{\emptyset, \{\emptyset\}, \{\{\emptyset\}\}, \{\emptyset, \{\emptyset\}\}\}$ is transitive by part 5 of Example 8.7. However, this set is **not** linearly ordered because $\{\{\emptyset\}\}$ and $\{\emptyset, \{\emptyset\}\}$ are incomparable (neither is an element of the other). Therefore, $\mathcal{P}(\{0,1\})$ is a transitive set that is **not** an ordinal.

38. Every set that is well-ordered by $\in$ is an ordinal.

False: For example, $\{0, 2\} = \{\emptyset, \{\emptyset, \{\emptyset\}\}\}$ is well-ordered by $\in$ because it is a subset of the well-ordered set $3 = \{0, 1, 2\}$. However, this set is **not** transitive because $\{\emptyset\} \in \{\emptyset, \{\emptyset\}\} \in \{\emptyset, \{\emptyset, \{\emptyset\}\}\}$, but $\{\emptyset\} \notin \{\emptyset, \{\emptyset, \{\emptyset\}\}\}$ (or equivalently, $1 \in 2 \in \{0, 2\}$, but $1 \notin \{0, 2\}$). Therefore, $\{0, 2\}$ is a set that is well-ordered by $\in$ that is **not** an ordinal.

39. If α, β, and γ are ordinals with $\alpha \in \beta$ and $\beta \in \gamma$, then $\alpha \in \gamma$.

True: Since γ is an ordinal, γ is transitive. Therefore, $\beta \in \gamma$ and $\alpha \in \beta$ implies that $\alpha \in \gamma$.

40. There exist two well-ordered sets such that neither one is isomorphic to an initial segment of the other one.

False: This follows from Well-ordered Set Fact 2.

LEVEL 3

Determine if each of the following ordinals is a cardinal. If so, explain why. If not, describe a bijection between this ordinal and a smaller ordinal. Use ordinal arithmetic where appropriate.

41. 0

0 is a cardinal because all natural numbers are cardinals.

42. 15

15 is a cardinal because all natural numbers are cardinals.

43. $\omega + 7$

$\boldsymbol{\omega + 7}$ **is not a cardinal**. We can define a bijection $f: \omega + 7 \to \omega$ by

$$f(\alpha) = \begin{cases} \alpha + 7 & \text{if } \alpha \in \omega. \\ k & \text{if } \alpha = \omega + k. \end{cases}$$

44. $7 + \omega$

$\boldsymbol{7 + \omega = \omega}$ **is a cardinal** because ω is the least infinite cardinal.

45. $\omega \cdot 7$

$\boldsymbol{\omega \cdot 7}$ **is not a cardinal**. We can define a bijection $f: \omega \cdot 7 \to \omega$ by

$$f(\alpha) = 7k + n, \text{ where } \alpha = \omega \cdot n + k.$$

46. 7ω

$\boldsymbol{7\omega = \omega}$ **is a cardinal** because ω is the least infinite cardinal.

47. ω^2

$\boldsymbol{\omega^2}$ **is not a cardinal**. Here is a visual representation of a bijection between ω and ω^2:

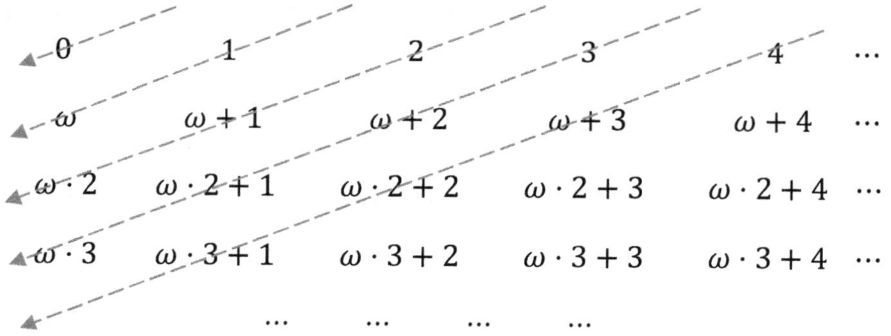

We can now list the elements of ω^2 as $0, 1, \omega, 2, \omega + 1, \omega \cdot 2, 3, \omega + 2, \omega \cdot 2 + 1, \omega \cdot 3, 4, \omega + 3, \ldots$

48. $5\omega^2 + 4$

$\boldsymbol{5\omega^2 + 4 = \omega^2 + 4}$ **is not a cardinal**. We can define a bijection $f: \omega^2 + 4 \to \omega$ by

$$f(\alpha) = \begin{cases} \alpha + 4 & \text{if } \alpha \in \omega^2. \\ k & \text{if } \alpha = \omega^2 + k. \end{cases}$$

Use ordinal arithmetic to write each of the following ordinals in the form $\omega^2 \cdot a + \omega \cdot b + c$, where $a, b \in \omega$.

49. $2\omega \cdot \omega$

$\omega \cdot \omega = \boldsymbol{\omega^2} = \boldsymbol{\omega^2 \cdot 1 + \omega \cdot 0 + 0}$

50. $(\omega \cdot \omega) \cdot 2$

$\boldsymbol{\omega^2 \cdot 2 = \omega^2 \cdot 2 + \omega \cdot 0 + 0}$

51. $\omega^2 \cdot 3 + 5\omega$

$\boldsymbol{\omega^2 \cdot 3 + \omega = \omega^2 \cdot 3 + \omega \cdot 1 + 0}$

52. $5\omega + 3\omega^2$

$\omega + \omega^2 = \boldsymbol{\omega^2} = \boldsymbol{\omega^2 \cdot 1 + \omega \cdot 0 + 0}$

53. $5\omega + \omega^2 \cdot 3$

$\omega + \omega^2 \cdot 3 = \boldsymbol{\omega^2 \cdot 3} = \boldsymbol{\omega^2 \cdot 3 + \omega \cdot 0 + 0}$

54. $\omega \cdot 5 + \omega^2 \cdot 3$

$\omega \cdot 5 + \omega^2 \cdot 3 = \boldsymbol{\omega^2 \cdot 3} = \boldsymbol{\omega^2 \cdot 3 + \omega \cdot 0 + 0}$

Let κ and λ be infinite cardinals. Verify each of the following:

55. $\kappa + \lambda = \max\{\kappa, \lambda\}$.

We may assume that $\lambda \leq \kappa$ (the argument is identical if we assume $\kappa \leq \lambda$).

Then $\max\{\kappa, \lambda\} = \kappa \leq \kappa + \lambda$.

Also, since $(\kappa \times \{0\}) \cup (\lambda \times \{1\}) \subseteq \kappa \times \lambda$, we have

$\kappa + \lambda = |(\kappa \times \{0\}) \cup (\lambda \times \{1\})| \leq |\kappa \times \lambda| = \kappa \cdot \lambda \leq \kappa \cdot \kappa = \kappa$ (by Cardinal Fact 1) $= \max\{\kappa, \lambda\}$.

Since $\kappa + \lambda \leq \max\{\kappa, \lambda\}$ and $\max\{\kappa, \lambda\} \leq \kappa + \lambda$, by the Cantor-Schroeder-Bernstein Theorem, we have $\kappa + \lambda = \max\{\kappa, \lambda\}$.

56. $\kappa \cdot \lambda = \max\{\kappa, \lambda\}$.

As in Problem 55, we may assume that $\lambda \leq \kappa$ (the argument is identical if we assume $\kappa \leq \lambda$).

Then $\max\{\kappa, \lambda\} = \kappa \leq |\kappa \times \lambda| = \kappa \cdot \lambda$.

Also, we have

$$\kappa \cdot \lambda = |\kappa \times \lambda| \leq |\kappa \times \kappa| = \kappa \cdot \kappa = \kappa \text{ (by Cardinal Fact 1)} = \max\{\kappa, \lambda\}.$$

Since $\kappa \cdot \lambda \leq \max\{\kappa, \lambda\}$ and $\max\{\kappa, \lambda\} \leq \kappa \cdot \lambda$, by the Cantor-Schroeder-Bernstein Theorem, we have $\kappa \cdot \lambda = \max\{\kappa, \lambda\}$.

57. $\kappa^3 = \kappa$.

By Cardinal Fact 1, $\kappa^2 = \kappa \cdot k = \kappa$. So, $\kappa^3 = \kappa^2 \cdot \kappa = \kappa \cdot \kappa = \kappa$, again by Cardinal Fact 1.

58. For each natural number $n \geq 1$, $\kappa^n = \kappa$.

We already know that $\kappa^n = \kappa$ for $n = 1, 2$, and 3 (by Part 4 of Example 8.21, Cardinal Fact 1, and Problem 53).

Now, suppose that $\kappa^j = \kappa$ for some natural number $j \geq 3$.Then $\kappa^{j+1} = \kappa^j \cdot \kappa^1 = \kappa \cdot \kappa = \kappa$.

LEVEL 4

Let κ, λ, and μ be infinite cardinals. Verify each of the following:

59. If $\mu \leq \lambda$, then $\kappa^\mu \leq \kappa^\lambda$.

The function $F: \kappa^\mu \to \kappa^\lambda$ defined by $F(f) = g$, where $g(\alpha) = f(\alpha)$ for all $\alpha \in \mu$ and $g(\alpha) = 0$ for all $\alpha \in \lambda \setminus \mu$ is injective.

60. $\kappa^\omega \leq \kappa^\lambda$.

Since λ is infinite, $\omega \leq \lambda$. So, the result follows from Problem 59.

61. $\kappa^\omega \leq \kappa^\mu$.

Since μ is infinite, $\omega \leq \mu$. So, the result follows from Problem 59.

62. If $\mu \leq \lambda$, then $\kappa^\lambda . \kappa^\mu = \kappa^\lambda$.

Since $\omega < 2^\omega \leq \kappa^\omega$, by Problems 60 and 61, κ^λ and κ^μ are both infinite. So, by Problem 56, we have $\kappa^\lambda \cdot \kappa^\mu = \max\{\kappa^\lambda, \kappa^\mu\} = \kappa^\lambda$. The last equality follows from Problem 59.

63. If $\mu \leq \lambda$, then $\kappa^{\lambda+\mu} = \kappa^\lambda$.

By Problem 55, we have $\lambda + \mu = \max\{\lambda, \mu\} = \lambda$. So, $\kappa^{\lambda+\mu} = \kappa^\lambda$.

64. $\kappa^\lambda \cdot \kappa^\mu = \kappa^{\lambda+\mu}$.

We may assume that $\mu \leq \lambda$ (the argument is identical if we assume $\lambda \leq \mu$).

By Problem 62, $\kappa^{\lambda}.\, \kappa^{\mu} = \kappa^{\lambda}$. By problem 63, $\kappa^{\lambda+\mu} = \kappa^{\lambda}$. Therefore, , $\kappa^{\lambda}.\, \kappa^{\mu} = \kappa^{\lambda+\mu}$.

Determine if each of the following statements is true or false. If true, explain why. If false, provide a counterexample.

65. If α is an ordinal, then $\in$ is a transitive relation on $\alpha + 1 = \alpha \cup \{\alpha\}$.

True

Let $\beta, \gamma, \delta \in \alpha + 1 = \alpha \cup \{\alpha\}$ with $\beta \in \gamma$ and $\gamma \in \delta$. If $\delta = \alpha$, then since α is a transitive set, $\beta \in \alpha$. So, $\beta \in \delta$. If $\delta \in \alpha$, then since α is a transitive set, $\gamma \in \alpha$, and again, since α is a transitive set, $\beta \in \alpha$. Since $\in$ is transitive on α, we have $\beta \in \delta$. So, $\in$ is a transitive relation on $\alpha + 1 = \alpha \cup \{\alpha\}$.

66. If α is an ordinal, then $\in$ is a trichotomous relation on $\alpha + 1 = \alpha \cup \{\alpha\}$.

True

Let $\beta, \gamma \in \alpha + 1 = \alpha \cup \{\alpha\}$ and assume that $\beta \notin \gamma$ and $\gamma \notin \beta$. If $\gamma = \alpha$, then since $\beta \notin \alpha$, we have $\beta = \alpha = \gamma$. Similarly, if $\beta = \alpha$, then since $\gamma \notin \alpha$, we have $\gamma = \alpha = \beta$. If $\beta \neq \alpha$ and $\gamma \neq \alpha$, then $\beta, \gamma \in \alpha$. Since $\in$ is trichotomous on α, we have $\beta = \gamma$. So, $\in$ is trichotomous on $\alpha + 1 = \alpha \cup \{\alpha\}$.

67. If α is an ordinal, then $\alpha + 1 = \alpha \cup \{\alpha\}$ is well-ordered by $\in$.

True

By Problem 65, $\in$ is transitive on $\alpha + 1 = \alpha \cup \{\alpha\}$.

By Problem 66, $\in$ is a trichotomous relation on $\alpha + 1 = \alpha \cup \{\alpha\}$.

Now, let $B \subseteq \alpha + 1 = \alpha \cup \{\alpha\}$ be nonempty.

If $B \setminus \{\alpha\} \neq \emptyset$, then $B \setminus \{\alpha\}$ is a nonempty subset of α, and therefore, has a least element β (because α is well-ordered by $\in$). Since $\beta \in \alpha$, β is the least element of B as well.

If $B \setminus \{\alpha\} = \emptyset$, then $B = \{\alpha\}$, and the least element of B is α. It follows that $\alpha + 1 = \alpha \cup \{\alpha\}$ is well- ordered by $\in$.

68. If α is a transitive set, then $\alpha + 1 = \alpha \cup \{\alpha\}$ is also a transitive set.

True

Let $\beta \in \alpha + 1 = \alpha \cup \{\alpha\}$ and let $\gamma \in \beta$. Then $\beta \in \alpha$ or $\beta \in \{\alpha\}$. If $\beta \in \alpha$, then γ is in α because α is an ordinal and therefore transitive. If $\beta \in \{\alpha\}$, then $\beta = \alpha$, and so, $\gamma \in \alpha$ because $\gamma \in \beta$.

69. If α is an ordinal, then $\alpha + 1 = \alpha \cup \{\alpha\}$ is also an ordinal.

True

This follows immediately from Problems 67 and 68.

70. If α is an ordinal and $x \in \alpha$, then x is an ordinal.

True

Let α be an ordinal and let $x \in \alpha$. Since α is transitive, $x \subseteq \alpha$. Since any subset of a well-ordered set is well-ordered, x is well-ordered by $\in$.

Now, suppose that $y \in x$ and $z \in y$. Since $x \subseteq \alpha$, we have $y \in \alpha$. Since α is transitive, $z \in \alpha$. Since $x, y, z \in \alpha$, $z \in y$, $y \in x$, and α is strictly linearly ordered by $\in$, we have $z \in x$. Therefore, x is transitive.

Since x is well-ordered by $\in$ and transitive, x is an ordinal.

Now, let $z \in x$. Since $x \subseteq \alpha$, $z \in \alpha$. Therefore, $z \in \text{pred}(\alpha, x)$. So, $x \subseteq \text{pred}(\alpha, x)$. Let $z \in \text{pred}(\alpha, x)$. Then $z \in \alpha$ and $z \in x$. In particular, $z \in x$. So, $\text{pred}(\alpha, x) \subseteq x$. Since we have $x \subseteq \text{pred}(\alpha, x)$ and $\text{pred}(\alpha, x) \subseteq x$, it follows that $x = \text{pred}(\alpha, x)$.

71. If κ a cardinal and $x \in \kappa$, then x is a cardinal.

False

ω_1 is a cardinal and $\omega + 1 \in \omega_1$, but $\omega + 1$ is **not** a cardinal.

72. If α is an ordinal and $x \in \alpha$, then $x = \text{pred}(\alpha, x)$.

True

Let $z \in x$. Since $x \subseteq \alpha$, we have $z \in \alpha$. Therefore, $z \in \text{pred}(\alpha, x)$. So, $x \subseteq \text{pred}(\alpha, x)$.

Now, let $z \in \text{pred}(\alpha, x)$. Then $z \in \alpha$ and $z \in x$. In particular, $z \in x$. So, $\text{pred}(\alpha, x) \subseteq x$.

Since we have $x \subseteq \text{pred}(\alpha, x)$ and $\text{pred}(\alpha, x) \subseteq x$, it follows that $x = \text{pred}(\alpha, x)$.

LEVEL 5

Let X be a nonempty set of ordinals. Verify each of the following:

73. $\bigcup X$ is a transitive set.

If $\alpha \in \bigcup X$, then there is $\beta \in X$ with $\alpha \in \beta$. Since β is an ordinal, β is transitive. Therefore, $\alpha \subseteq \beta$. Since $\beta \subseteq \bigcup X$, $\alpha \subseteq \bigcup X$. Since $\alpha \in \bigcup X$ was arbitrary, $\bigcup X$ is transitive.

74. $\bigcup X$ is well-ordered by $\in$.

Let $\beta, \gamma, \delta \in \bigcup X$ with $\beta \in \gamma$ and $\gamma \in \delta$. By the definition of the union, there are $x, y, z \in X$ with $\beta \in x$, $\gamma \in y$, and $\delta \in z$. By the definition of X, x, y, and z are ordinals. By the solution to Problem 70 above, β, γ, and δ are ordinals. By Problem 39, $\beta \in \delta$. So, $\in$ is transitive on $\bigcup X$.

Next, let $\beta, \gamma \in \bigcup X$. As in the last paragraph, there are ordinals $x, y \in X$ with $\beta \in x$ and $\gamma \in y$. We may assume that $x \leq y$ (that is $x \in y$ or $x = y$). Therefore, since y is transitive, $\beta \in y$. Since $\beta, \gamma \in y$ and y is an ordinal, we must have $\beta \in \gamma$, $\beta = \gamma$, or $\gamma \in \beta$. So, $\in$ is trichotomous on $\bigcup X$.

Finally, let Y be a nonempty subset of $\bigcup X$ and let α be any element of Y. By the definition of union, $\alpha \in X$, and so, α is an ordinal. In particular, α is well-ordered by $\in$. If $\alpha \cap Y \neq \emptyset$, then it has a least element, which is also the least element of Y. Otherwise, α is the leaset element of Y.

It follows that $\bigcup X$ is well-ordered by $\in$.

75. If $\alpha \in X$, then $\alpha \leq \bigcup X$.

Suppose $\alpha \in X$. Then $\alpha \subseteq \bigcup X$. Since α and $\bigcup X$ are both ordinals, we must have $\alpha \in \bigcup X$, $\alpha = \bigcup X$, or $\bigcup X \in \alpha$. Now, if $\bigcup X \in \alpha$, then since $\alpha \subseteq \bigcup X$, $\bigcup X \in \bigcup X$, which is impossible. So, $\alpha \in \bigcup X$ or $\alpha = \bigcup X$. That is $\alpha \leq \bigcup X$.

76. $\bigcup X$ is the least ordinal greater than or equal to all elements of X.

By Problems 73 and 74, $\bigcup X$ is an ordinal.

By Problem 75, $\bigcup X$ is greater than or equal to all elements of X.

Now, suppose that there is an ordinal β such that $\alpha \in \beta$ for all $\alpha \in X$. Let $\gamma \in \bigcup X$. Then there is $\alpha \in X$ with $\gamma \in \alpha$. Since $\alpha \in X$, we have $\alpha \in \beta$. By the transitivity of β, $\gamma \in \beta$. Since $\gamma \in \bigcup X$ was arbitrary, $\bigcup X \subseteq \beta$. We cannot have $\beta \in \bigcup X$, because then we would have $\beta \in \beta$. Therefore, $\bigcup X \leq \beta$. It follows that $\bigcup X$ is the **least** ordinal greater than or equal to all elements of X.

77. $\bigcap X$ is an ordinal.

If $\bigcap X = \emptyset$, then $\bigcap X = 0$, which is an ordinal. So, we may assume that $\bigcap X \neq \emptyset$.

We first show that $\bigcap X$ is transitive.

Let $\alpha \in \bigcap X$ and $\beta \in \alpha$. If $\gamma \in X$, then we have $\beta \in \alpha \in \gamma$. So, $\beta \in \gamma$ for all $\gamma \in X$. Therefore, $\beta \in \bigcap X$. Since $\beta \in \alpha$ was arbitrary, $\alpha \subseteq \bigcap X$. Since $\alpha \in \bigcap X$ was arbitrary, $\bigcap X$ is transitive.

Now, let $\beta, \gamma, \delta \in \bigcap X$ with $\beta \in \gamma$ and $\gamma \in \delta$. Let x be any element of X. Then $\beta, \gamma, \delta \in x$. By the definition of X, x is an ordinal. By the solution to Problem 70 above, β, γ, and δ are ordinals. By Problem 39, $\beta \in \delta$. So, $\in$ is transitive on $\bigcup X$.

Next, let $\beta, \gamma \in \bigcap X$. As in the last paragraph, let x be some ordinal in X. Then with $\beta, \gamma \in x$. We may assume that $x \leq y$ (that is $x \in y$ or $x = y$). Therefore, since y is transitive, $\beta \in y$. Since $\beta, \gamma \in y$ and y is an ordinal, we must have $\beta \in \gamma$, $\beta = \gamma$, or $\gamma \in \beta$. So, $\in$ is trichotomous on $\bigcup X$.

Finally, let Y be a nonempty subset of $\bigcap X$ and let α be any element of X. Then since $\bigcap X \subseteq \alpha$, we have $Y \subseteq \alpha$. Since α is an ordinal, Y has a least element.

It follows that $\bigcap X$ is well-ordered by $\in$.

78. $\bigcap X$ is the least ordinal in X.

Let α be the least ordinal in X. Let $\beta \in \alpha$ and $\gamma \in X$. Since α is the least ordinal in X, $\gamma = \alpha$ or $\alpha \in \gamma$. In either case, $\beta \in \gamma$. So, $\beta \in \gamma$ for every $\gamma \in X$. Therefore, $\beta \in \bigcap X$. Since $\beta \in \alpha$ was arbitrary, $\alpha \subseteq \bigcap X$. Now, let $\beta \in \bigcap X$. Since $\alpha \in X$, $\beta \in \alpha$. Since $\beta \in \bigcap X$ was arbitrary, $\bigcap X \subseteq \alpha$. Since $\alpha \subseteq \bigcap X$ and $\bigcap X \subseteq \alpha$, we have $\bigcap X = \alpha$. Therefore, $\bigcap X$ is the least ordinal in X.

79. If every ordinal in X is a cardinal, then $\bigcup X$ is a cardinal.

By Problem 76, $\bigcup X$ is an ordinal. Suppose toward contradiction that $|\bigcup X| \in \bigcup X$. Then there is $\kappa \in X$ with $|\bigcup X| \in \kappa$. Again, by Problem 76, $\kappa \in \bigcup X$. Therefore, $\kappa = |\kappa| \leq |\bigcup X|$. So, we have $\kappa \in \kappa$, which is impossible. Therefore, $|\bigcup X| = \bigcup X$. So, $\bigcup X$ is a cardinal.

Determine if each of the following statements is true or false. If true, explain why. If false, provide a counterexample.

80. Every infinite cardinal is a limit ordinal.

True

Every infinite successor ordinal is equinumerous with its predecessor. In other words, if $\alpha + 1$ is infinite, then $|\alpha + 1| = |\alpha|$, and therefore, $\alpha + 1$ is not a cardinal. It follows that every infinite cardinal is a limit ordinal.

81. Every infinite successor cardinal is regular.

True

Let κ be an infinite successor cardinal. Then $\kappa = \alpha^+$ for some ordinal α. Suppose toward contradiction that $cf(\kappa) < \kappa$. Then there is an ordinal $\beta < \kappa$ and a cofinal map $f\colon \beta \to \kappa$. It follows that $\kappa = \bigcup\{f(\gamma) \mid \gamma \in \beta\}$. Now, for each $\gamma \in \beta$, $|f(\gamma)| < \kappa$. So, $|f(\gamma)| \leq |\alpha|$. Also, since $\beta < \kappa$, we have $|\beta| < \kappa$. So, $|\beta| \leq |\alpha|$. We have $\kappa = |\bigcup\{f(\gamma) \mid \gamma \in \beta\}| \leq |\alpha| \leq \alpha < \kappa$, a contradiction.

82. Every limit cardinal is regular.

False

$cf(\omega_\omega) = \omega$. Therefore, ω_ω is a limit cardinal that is not regular.

About the Author

Dr. Steve Warner, a New York native, earned his Ph.D. at Rutgers University in Pure Mathematics in May 2001. While a graduate student, Dr. Warner won the TA Teaching Excellence Award.

After Rutgers, Dr. Warner joined the Penn State Mathematics Department as an Assistant Professor and in September 2002, he returned to New York to accept an Assistant Professor position at Hofstra University. By September 2007, Dr. Warner had received tenure and was promoted to Associate Professor. He has taught undergraduate and graduate courses in Precalculus, Calculus, Linear Algebra, Differential Equations, Mathematical Logic, Set Theory, and Abstract Algebra.

From 2003 – 2008, Dr. Warner participated in a five-year NSF grant, "The MSTP Project," to study and improve mathematics and science curriculum in poorly performing junior high schools. He also published several articles in scholarly journals, specifically on Mathematical Logic.

Dr. Warner has nearly two decades of experience in general math tutoring and tutoring for standardized tests such as the SAT, ACT, GRE, GMAT, and AP Calculus exams. He has tutored students both individually and in group settings.

In February 2010 Dr. Warner released his first SAT prep book "The 32 Most Effective SAT Math Strategies," and in 2012 founded Get 800 Test Prep. Since then Dr. Warner has written books for the SAT, ACT, SAT Math Subject Tests, AP Calculus exams, and GRE. In 2018 Dr. Warner released his first pure math book called "Pure Mathematics for Beginners." Since then he has released several more books, each one addressing a specific subject in pure mathematics.

Dr. Steve Warner can be reached at

steve@SATPrepGet800.com

BOOKS BY DR. STEVE WARNER

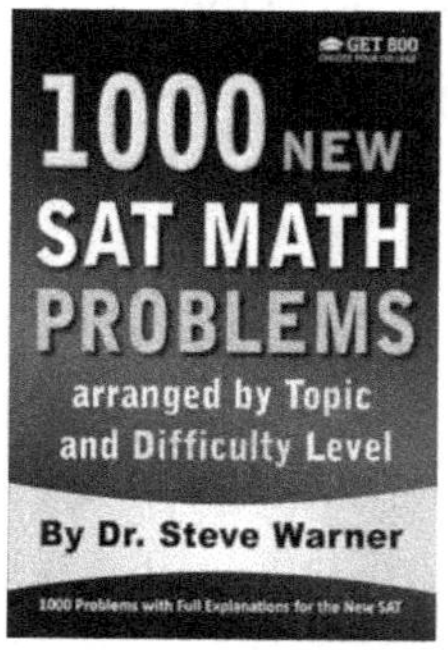

GET 800
PHYSICS MASTERY
for the Advanced High School Student
Complete Physics Review with 400 SAT and AP Physics Questions

GET 800
320
GRE MATH PROBLEMS
arranged by Topic and Difficulty Level
By Dr. Steve Warner

Second Edition
320
AP CALCULUS AB PROBLEMS
arranged by Topic and Difficulty Level
By Dr. Steve Warner

GET 800
Second Edition
320
AP CALCULUS BC PROBLEMS
arranged by Topic and Difficulty Level
By Dr. Steve Warner

GET 800
320
SAT CHEMISTRY SUBJECT TEST PROBLEMS
arranged by Topic and Difficulty Level
Christopher Bozza
Dr. Steve Warner

GET 800
NEW
SHSAT
Verbal Prep Book To Improve Your Score In Two Months
The Most Effective Strategies for Mastering Reading Comprehension and Revising/Editing on the SHSAT
By Larry Ronaldson M.S. and Deirdre Storck B.A.

1000 Logic and Reasoning Questions for
GIFTED AND TALENTED
Elementary School Students

GET 800
Vocabulary Builder:
The Complete Vocabulary Word Workshop for Building English Vocabulary
BONUS INSIDE!

Algebra Handbook
for Gifted Middle School Students
Dr. Steve Warner
Tayyip Oral
Serife Turan

555 Geometry Problems
for High School Students
Dr. Steve Warner
Tayyip Oral
Serife Turan

555 Math IQ Questions
for Middle School Students
Dr. Steve Warner
Tayyip Oral

555 Advanced Math Problems
for Middle School Students
Dr. Steve Warner
Tayyip Oral

www.ingramcontent.com/pod-product-compliance
Lightning Source LLC
Chambersburg PA
CBHW081746200326
41597CB00024B/4410
* 9 7 8 1 9 5 1 6 1 9 0 5 3 *